1 MONTH OF
FREE
READING

at

www.ForgottenBooks.com

By purchasing this book you are eligible for one month membership to ForgottenBooks.com, giving you unlimited access to our entire collection of over 700,000 titles via our web site and mobile apps.

To claim your free month visit:
www.forgottenbooks.com/free465498

ISBN 978-0-266-62669-5
PIBN 10465498

VEGETABLE TECHNOLOGY:

CONTRIBUTION TOWARDS A BIBLIOGRAPHY OF ECONOMIC
BOTANY, WITH A COMPREHENSIVE
SUBJECT-INDEX.

BY

BENJAMIN DAYDON JACKSON,

SECRETARY OF THE LINNEAN SOCIETY.

FOUNDED UPON THE COLLECTIONS OF
GEORGE JAMES SYMONS, F.R.S.

LONDON:
PUBLISHED FOR THE INDEX SOCIETY
By LONGMANS, GREEN & Co., 39, PATERNOSTER ROW,
AND DULAU & Co., 37, SOHO SQUARE.
MDCCCLXXXII.

HERTFORD:

PRINTED BY STEPHEN AUSTIN AND SONS.

CONTENTS.

PREFACE.

THE history of this volume is as follows. In the *Colonies and India* for Sept. 13th, 1879, Mr. G. J. Symons published a catalogue of works on Applied Botany, having previously contributed a letter to that periodical urging the compilation of such a catalogue. The same journal for November 22nd, 1879, contained an appendix consisting of the former list, with large accessions by Mr. P. L. Simmonds. Mr. Symons having added a few additional titles to the list, it was offered to the Index Society for publication, and I undertook to edit it. The work served excellently as a supplement to a journal, but required much labour before it could be considered fit for publication in a permanent form. I began by striking out colonial floras of purely botanical aspect, they having been far more fully dealt with in my *Guide to the Literature of Botany.* I also deleted all books devoted to Silk and Cochineal, as not being strictly within the limits of the title as settled by the Index Society; and most reluctantly the subject of the Vine, its culture and products, simply on the ground of its enormous extent. The bibliography of the Vine in all its bearings would require a lifetime for its compilation; I did not attempt to take up the subject myself: in the Index I have made some references to works which will help inquirers on their way, but the nineteen treatises which were cited in the original list were many of them only of gardening interest. I have excluded books and papers of simply horticultural, therapeutic, chemical, commercial, or manufacturing interests, unless they contained a sufficient account of the raw product, its cultivation, or whence obtained, to justify their citation.

Mr. Symons supplied me with about sixty cuttings from catalogues not incorporated with his lists. I checked these where possible, against the volumes themselves either at the British Museum, Linnean Society, Kew Museum, or elsewhere; if I could not find any trustworthy record, I rejected them unhesitatingly. I was compelled to examine *every* title without exception, for too great trust had apparently been placed on the catalogues whence they were obtained for me to reprint them. Many of the original entries were too imperfect to be retained; those which I could not supplement or correct were mostly struck out; a few, however, are still left in, but marked with an asterisk, to signify that I am in no wise responsible for their accuracy. I cannot here refrain from expressing my great regret that Mr. Simmonds should have contented himself with giving so many maimed entries from pamphlets in his own possession under the heading 'Anonymous' publications; a trifling amount of additional trouble would have made them quotable or recognizable.

In addition to examining the titles furnished me, I have added all I could conveniently within the time; but a complete bibliography of Economic Botany would need the labour of years. I have gone through the Library attached to the

Economic Museum at Kew, the *Bibliotheca Historico-Naturalis* from 1862 to 1880, the *Botanisches Centralblatt* since its establishment last year to the present time, *The Pharmaceutical Journal* from its formation in 1842 to the end of October last. Mr. H. B. Wheatley was kind enough to furnish me with the titles of articles in the *Journal of the Society of Arts* to the end of 1880, since then I have taken them out myself. Taken altogether, these entries supply a very fair starting-point for almost every question in Vegetable Technology; the number of articles brought forward from the original list is 326, which has been increased, as noted, to 3580, exclusive of translations and different editions. One peculiarity of the old list was its paucity of works in the German language; this has been to some extent remedied in the present volume, as a book which professed to treat of forestry and ignored German literature on the subject, would be sadly deficient.

The *Journal of Applied Science* and *The Technologist* have not been regularly searched and cited, like the *Journal of the Society of Arts* and *The Pharmaceutical Journal;* for whilst confidence may be reposed in the statements given in these two latter journals, I could not confidently assert the same of the two former.

I have endeavoured to meet the wants of inquirers by compiling a comprehensive index. This not being an elementary work on applied botany, I have not attempted the task of determining the various products which appear under the native names, sometimes in different spellings; to have done so critically would have consumed months to very little purpose. The Index is designed to help in finding the books on a given subject, which are distributed under the authors' names, and more than this should not be demanded; a directory does this

for persons, without trying to define the mutual affinities of
those bearing the same name. It is hardly necessary to insist
upon the need of searching all entries likely to give the
information sought; thus, if an inquirer do not find his
wants satisfied when referring to 'Materia Medica,' he should
look up cognate headings, such as 'Drugs' and 'Medicinal
Plants.'

I should have been glad to include a complete series of
references to consular reports, also the selections of Indian
papers bearing on Vegetable Technology, but I found I could
not attain even approximate completeness therein. Consular
reports are often quoted in *The Pharmaceutical Journal*, whence
they are cited here, and a good series of the Indian selections is
to be found in the Library Catalogue of the Royal Geographical
Society. These publications are often regarded as ephemeral,
and the department responsible for their issue rarely has a
complete set.

Parliamentary papers have not been quoted in full, for in
the majority of cases their value is commercial or manufacturing; in the Index, however, I have made references to
suggest paths which may be followed up by those who want
more than I have given. The formulae '*Refer to*' and '*Refer
also to*' signify something outside this volume, whilst '*See*'
and '*See also*' are ordinary cross-references.

The rules of the Index Society are followed in their spirit,
the article preceding a name being used in the alphabetical
arrangement, so that 'De Vrij,' and similar Dutch names,
figure under the article, with a cross-reference from the sub-
stantive following. As in the *Guide to the Literature of
Botany*, modified vowels and diphthongs are spelled out, ä, ö, ü,
becoming ae, oe, ue, and so forth.

The heading 'Anonymous Publications' has been much altered from its original shape. In the former list there were several entries which have been here reduced to their proper authorship; others have been excluded by the change of plan; many were too hopelessly defective to be here given. The system of cataloguing anonymous books at the British Museum is an excellent method of drafting off such ware into the Library, but it gives almost no assistance to the searcher. In some cases it is not enough to know all that the book itself states, as where a medical tract by John Pechie is entered under Sir John Micklethwaite's name, *because*, Pechie addressed his pamphlet " To the President of the College of Phisitians," and the cataloguer having taken the pains to find out who was President of the College for the time being, entered it under his name. The Old General Catalogue, with its common-sense entries, is far more useful for hunting up these books than the New General Catalogue, with its elaborate system of rules.

The abbreviations used will, I believe, in every case be recognized without giving a tabular statement—'*Pharm. Journ.*' for *Pharmaceutical Journal*, and so on. In this, as in Mr. Solly's *Index to . . . Titles of Honour*, and my *Guide to the Literature of Botany*, the sign $\rightarrow$ is used to signify 'in progress'; many of the abbreviations I was compelled to use in my *Guide to the Literature of Botany* are not used here, as I have been able to quote the titles fully, any omissions in the middle being shown by dots . . . , at the end by *etc.*

During the progress of the work I have been indebted to many friends for help on special points, for which I here tender my hearty thanks; but I must especially name Mr. C. G. Warnford Lock for much help afforded throughout, and for supplying many titles of papers and books which I should otherwise have missed.

In spite of inevitable shortcomings, I believe that this volume will prove to be of greater use than its predecessors on the same topics. Dryander's Catalogue of the Banksian Library, which is arranged most minutely as to subject, and the systematic portion in both editions of Pritzel's *Thesaurus*, are less convenient for reference than the one alphabetic index which closes this book.

B. DAYDON JACKSON.

30, STOCKWELL ROAD, LONDON, S.W.,

    27th December, 1881.

VEGETABLE TECHNOLOGY.

PART I.

CATALOGUE OF AUTHORS.

ABBAY (Richard).
Observations on Hemileia vastatrix, the so-called coffee-leaf disease. *Journ. Linn. Soc., Botany,* xvii. (1878) 173–184.

ABL (*Dr. —.*).
On the use of Coca leaves. [From *The Technologist.*] *Pharm. Journ.* II. vii. (1865) 33–34.

ABLETT (William H.).
English Trees and Tree-planting. London, 1880. 8°.

ACCUM (Frederick).
A Treatise on adulterations of food, and culinary poisons, exhibiting the fraudulent sophistications of . . . articles employed in domestic economy, and methods of detecting them. London, 1820. 12°. Ed. 2. 1820. 8°.

ACHARD (Franz Carl).
Ausfuehrliche Beschreibung der Methode, nach welcher bei der Kultur der Runkelruebe verfahren werden muss. Berlin, 1799. 8°.

Anleitung zur Bereitung des Rohzuckers und des Syrups aus den Runkelrueben. Berlin, 1800. 8°.

Beantwortung der Frage : wie ist die Zuckerfabrication aus Runkelrueben und des Branntweins aus den dabei abfallenden Abgaengen in den Preuss. Staaten zu bearbeiten, damit die koenigl. Accis-Gefaelle nicht dadurch bedenkliche Ausfaelle leiden? Berlin, 1800. 8°.

Kurze Geschichte der Beweise, welche ich von der Ausfuehrbarkeit im Grossen und den vielen Vortheilen der Zuckerfabrikation aus Runkelrueben gefuehrt habe. Berlin, 1800. 8°.

Anbau der zur Fabrikation anwendbaren Runkelrueben, und zur Gewinning des Zuckers aus denselben. Breslau, 1803. 8°.

Nachricht ueber die Runkelruebenzuckerfabrikation zu Kunern in Schlesien, welcher beglaubte Proben der Haupt- und Neben-Fabrikate, *etc.* Breslau, 1805. 8°.

ACHARD (Franz Carl), *continued :—*

Ueber den Einfluss der Runkelrueben-Zuckerfabrikation auf die Oekonomie. Glogau, 1805. 8°.

Die europaeische Zuckerfabrikation aus Runkelrueben, in Verbind-ung mit der Bereitung des Branntweins, des Rums, des Essigs und eines Kaffee-Surragats aus ihren Abfaellen. Leipzig, 1809. 8°. Ed. 2. 1812.

———— [Transl. by Dr. —. Angar.] Traité complet sur le sucre européen de betteraves, culture de cette plante considérée sous le rapport agronomique et manufacturier. Paris, 1812. 8°.

Die Zucker- und Syrup-Fabrikation aus Runkelrueben, als ein von jedem Gutsbesitzer mit Vortheil auszufuehrender Neben-zweig des oekonomischen Erwerbes. Breslau, 1810, 8°. Ed. 2. 1812.

———— [Transl. by Dr. —. Angar.] Instruction sur la fabrication du sucre et sirop de betterave. Paris, 1811. 8°.

ADRIANI (A.).

Ou Gutta Percha, Caoutchouc, and the Milky Juice of Ficus elastica. [From the *Central Blatt*.] *Pharm. Journ.* x. (1851) 546–549.

AGNEW (Ernest T.).

The Economic Uses of Malvaceae. *Pharm. Journ.* III. ii. (1871) 82.

AGUADO (Ramon Romualdo).

Tratado del arbolista teórico y práctico, . . . y una relacion por órden alfabético de las plantas arbóreas que mas abundan en los cultivos europeas y en nuestra peninsula. Madrid, 1864. 8°.

AGUIAR (J. M. de).

The botanical source of Araroba. [Andira Araroba, *Aguiar.* Abstract.] *Pharm. Journ.* III. x. (1879) 42–44.

AHLES (W.).

Unsere wichtigeren Giftgewaechse mit ihren pflanzlichen Zerglie-derung und erlaeut. Text, *etc.* 3. . . . Auflage von. M. Ch. F. Hochstetter's Giftgewaechse Deutschlands und der Schweiz. Esslingen, 1874–6. fol.

AINSLIE (Whitelaw).

Materia medica of Hindostan, and artisan's and agriculturist's nomenclature. Madras, 1813. 4°.

Materia indica; or some account of those articles which are em-ployed by the Hindoos, and other eastern nations, in their medicine, arts, and agriculture, *etc.* London, 1826. 2 vols. 8°.

AITCHISON (James Edward Tierney).

Hand-book of the Trade Products of Leh, *etc.* Calcutta, 1874. 8°.

ALBERTI (R.)

Zweiter und dritter Bericht ueber die Thaetigkeit der Versuchs-station des land- und forstwirthschaftlichen Provinzial-Vereins fuer das Furstenth. Hildesheim. Celle, 1875. 8°.

ALBERTUS [DE BOLLSTAEDT, *styled*] MAGNUS.

Ausfuehrliches Kraeuterbüch, oder gruendliche Beschreibung aller heilwirkenden Pflanzen, Kraeuter und Gestraeuche mit genauer Angabe ihrer Verwendung, Kraefte und Wirkungen. Reutlingen, 1871. 16°.

ALEFELD (Friedrich, *né* LECHDRINGHAUSEN).

Die Bienen-Flora Deutschlands und der Schweiz. Darmstadt, 1856. 8°. Ed. 2. Neuwied, 1863.

Landwirthschaftliche Flora, oder die nutzbaren cultivirten Garten- und Feldgewaechse Mitteleuropas in allen ihren wilden und Culturvarietaeten. Berlin, 1866. 8°.

ALFONSO (Ferdinando).

Trattato sulla coltivazione degli agrumi. Ed. 2. Palermo, 1875. 8°.

Monografia sui tabacchi della Sicilia. Palermo, 1880. 8°.

ALISON (S. Scott).

An account of a Cucurbitaceous fruit, employed as a violent purgative by the natives of Brazil, and recently brought to England, *etc.* [Luffa sp.] *Pharm. Journ.* iv. (1845) 360–362.

ALLART (F. A.).

Traité de la culture du tabac, indiquant tous les moyens à employer, depuis la disposition du terreau pour les couches, jusqu'à la mise en entrepôt de la récolte. Abbeville, 1876. 8°.

ALLEN (Charles B.).

Note on the history of Saffron. *Pharm. Journ.* III. xi. (1880), 449–450. *See also* pp. 461–463.

ALLEN (S. Stafford).

On the cultivation of Egyptian Opium. *Pharm. Journ.* II. iv. (1862) 199.

ALOI (Antonio).

Manuale teorico-pratico per la coltivazione dell' ulivo. Milano, 1875. 16°.

Forms vol. xxviii. of the ' Biblioteca dell' agricoltore.'

AMERY (C. F.).

Notes on Forestry. London, 1875. 8°.

AMOREUX (Pierre Joseph).

Traité de l'olivier, contenant l'histoire et culture de cet arbre, *etc.* Ed. 2. Montpellier, 1784. 8°.

> Ed. 1. was issued in the Recueil de l'academie, with other essays by Bernard and Coutoure.

ANDERSON (A. W.).

Notes on the Sugar-cane and the manufacture of Sugar in the West Indies. Trinidad, 1860. 8°.

ANDERSON (G.).

Jottings on Coffee Culture in Mysore. Bangalore, 1879. 8°.

ANDERSON (Thomas).

First annual report on the experimental cultivation of the Quiniferous Cinchona in British Sikkim, from 1st April, 1862, to 30th April, 1863. [From *The Calcutta Gazette.*] *Pharm. Journ.* II. v. (1863) 222–226.

Analysis of Darjeeling Chinchona Bark. *Pharm. Journ.* II. ix. (1867) 244–245.

ANDEREGG (F.).

Der Tabakbau in der Schweiz. Chur, 1880. 8°.

ANDREWS (Charles).

Note on Areca nut. *Pharm. Journ.* III. iv. (1873) 649.

ANDREWS (Michael).

Instructions for the culture and preparation of flax in Ireland. 9th issue. Belfast, 1880. 8°.

ANDY (S. Pulney).

On the use of Margosa [Azadirachta indica] leaves in smallpox. [From *The Madras Quarterly Journal.*] *Pharm. Journ.* II. ix. (1868) 392–393.

ANONYMOUS WORKS. *See end of Alphabetical list of Authors.*

ANSBERQUE (Edmonde).

Flore fourragére de la France, reproduite par méthode de compression, dite phytoxygraphique, et publiée sous le patronage de la ville de Lyon. Lyon, 1866. fol.

ANSLIJN (Nicolaas).

Afbeelding der artzenij-gewassen. [Nos. 1–54.] Amsterdam, 1829–39. fol.

Handleiding in de leer der botanic, . . . eene sijstematische beschrijving dier gewassen, welke in de Nederlandsche apotheek vermeldzijn. Amsterdam, 1831. 8°.

Handleiding tot de kennis der artzenii-gewassen, welke in de Nederlandsche apotheek zijn obgenomen, voornamelijk ingerigt ten behoeve van hen, die zich der geneescheel en artzenijmengkunde wijdende, *etc.* Leijden, 1835–38. 8°.

Antelme (C.).

Mémoire sur la culture de la canne à sucre à l'île Maurice. Bordeaux, 1866. 8°.

Antisell (Thomas).

Cundurango. *Pharm. Journ.* III. ii. (1871) 62–63.

Anton (C.).

Die Giftgewaechse Deutschlands, Oesterreichs und der Schweiz. Nebst Angabe der sie kennzeichn. Merkmale, *etc.* Neu-Ulm, 1879. 8°.

Antz (Karl Caesar).

Tabaci historia. Berolini, 1836. 8°.

Arata (Pedro N.).

The "Gum" of the Quebracho colorado (Loxopterigium Lorentii, Grisebach). [Extract.] *Pharm. Journ.* III. ix. (1878) 531.

Arbo (A.).

Compendium i medicinisk Botanik. 2den Udgave, omarbeidet ved N. Bryhm, *etc.* Christiania, 1875. 8°.

Archer (Thomas Croxen).

Lecture on the starch-producing plants of commerce. *Pharm. Journ.* xi. (1852) 556–557.

Popular Economic Botany, or description of the botanical and commercial characters of the principal articles of vegetable origin, used for food, clothing, tanning, dyeing, building, medicine, perfumery, etc. London, 1853. 8°. Ed. 2. 1876.

First steps in Economic Botany, for the use of students; being an abridgment of Popular Economic Botany. London, 1854. 16°.

A brief notice of a few articles imported into Liverpool during 1853. *Pharm. Journ.* xiii. (1854) 312–313.

Second notice of new or rare articles imported into Liverpool. *Pharm. Journ.* 447–449.

The Ordeal Bean of Old Calabar. *Pharm. Journ.* xiv. (1855) 525–526.

The useful products of the natural order Palmaceae. [Abstract.] *Pharm. Journ.* xv. (1855) 263–265.

On some of the Animal and Vegetable Products Constituting the Foreign Commerce of Liverpool. *Journ. Soc. Arts,* iv. (1856) 255–258.

The useful products of the natural order Graminaceae. [Abstract.] *Pharm. Journ.* xv. (1856) 548–549.

The new styptic, Penghawar Djambi. [Cibotium sp.] *Pharm. Journ.* xvi. (1856) 322–323.

Vegetable products of the world in common use. London, 1862. 8°.

Archer (Thomas Croxen), *continued :*—
Some account of Paullinia sorbilis and its products. [Extract.]
Pharm. Journ. II. v. (1863) 135–136.
Profitable plants. London, 1865. 8°.
A reissue of Popular Economic Botany, with new title-page.

Arduino (Pietro).
Memorie di osservazioni e di sperienze sopra la coltura e gli usi di
varie piante, che servono o che servir possono utilmente alla
tintura, all' economia, all' agricultura, *etc.* Tomo i. Padova,
1766. 4°.
Del genera delle avene, delle sua specie e varietà, della coltura ed
usi economici. Padova, 1789. 4°.

Argenti (Vicente Martin de).
Album de la flora médico-farmacéutica e industrial indígena y
exótica ó sea coleccion de laminas iluminadas de las plantas
de aplicacion en la medicina, farmàcia, industria, y artes, *etc.*
Madrid, 1866. 3 vols. fol.

Arioli (V. E.).
Prontuario delle piante medicinali e industriali, a commodo dei
botanici, farmacisti, medici e infermieri ; con appendice.
Milano, 1878. 32°.

Arnot (William).
The Technology of the Paper Trade. *Journ. Soc. Arts.* xxvi.
(1877) 59–65, 73–78, (1878) 89–95, 101–107, 147–153, 164–169.
Also published separately as Cantor Lectures.

Artus (Wilhelm Friedrich Wilibald).
Atlas aller in den neuesten Pharmacopoen Deutschlands aufgeno-
menenen officinellen Gewaechse . . . Zugleich ein Hilfs- und
Ergaenzungswerk aller bisherigen Pharmacopoeen, pharmacognostischen und pharmacologischen . Werke. Leipzig,
1864–67. 4°.

Ascherson (Ferdinand Moritz).
Pharmaceutische Botanik in Tabellenform. Berlin, 1831. 4°.

Aschieri (G.).
Dizionario di scienza organico-vegetale di agricoltura, Fasc. 1.
Milano, 1863. 8°.

Ashworth (Henry).
Cotton; its cultivation, manufacture, and uses. Manchester, 1858.
8°. Also in *Journ. Soc. Arts,* vi. (1858) 256–264.

Attfield (John).
On the nutritive value of Dika bread. *Pharm. Journ.* II. ii.
(1862) 445–447.

ATIFIELD (John), *continued :—*

A contribution to the history of Balsam of Peru. *Pharm. Journ.* II. vi. (1864) 204–206.

On the food-value of the Kola-nut—a new source of Theine. [See also DANIELL.] *Pharm. Journ.* II. vi. (1865) 457–460.

Analysis of Eland's Boontjes; a species of Acacia yielding food, medicine, and tan to the natives of South Africa. *Pharm. Journ.* II. viii. (1866) 316–318.

Note on mustard oil-cake. *Pharm. Journ.* II. x. (1869) 520–521.

ATKINSON (Edwin T.).

Gums and Gum Resins. Allahabad, 1876. 8°.
This is part 1. of Economic products of the North-Western Provinces.

AUBLET (Jean Baptiste Christophore Fusée).

Histoire des plantes de la Guiane française . . . relatifs à la culture et au commerce de la Guiane française, *etc.* Londres et Paris, 1775. 4 vòls. 4°.

AVEQUIN (J. B.).

Maple Sugar of the United States. [From the *Journ. de Pharm.*] *Pharm. Journ.* xvii. (1857) 324–327.

BABO (A. von).

Traité de la culture du tabac. Traduite de l'allemand par A. Dauphiné. Paris, 1861. 8°.

BADHAM (Charles David).

The esculent mushrooms of England. London, 1847. 8°. Ed. 2. by F. Currey, 1864.

BAEUMKER (J.).

Experimentelle Beitraege zur Kenntniss der pharmakologischen Wirking der Frangularinde. Goettingen, 1880. 8°.

BAGNERIS (G.).

Manuel de sylviculture. Nancy, 1873. 12°.

BAGUENAULT DE VIÉVILLE (—.).

Observations pratiques sur la culture et l'aménagement des pins maritimes et sylvestres dans la Sordogne centrale. Orléans, 1875. 8°.

Le Chêne. Orléans, [1876]. 8°.

BAILEY (—.) & Co.

Forest Culture : its Value, Profits, *etc.* San Francisco, 1875. 12°.

BAILDON (Henry Craven).

Brief remarks on the Bark of Rhamnus Frangula, or Black Alder Tree, a shrub of the north of Europe. *Pharm. Journ.* III. ii. (1871) 152.

Cortex Rhamni Frangulae. *Pharm. Journ.* III. iv. (1874) 889.

BAILDON (Samuel).

Tea in Assam. A pamphlet on the origin, culture, and manufacture of Tea in Assam, *etc.* Calcutta, 1877. 8°.

BAILLON (Henri Ernest).

The botanical origin and characters of the officinal Rhubarbs.
[Extract.] *Pharm. Journ.* III. iii. (1872) 301.

Botanical origin of the Balsams of Tolu and Peru. *Pharm.
Journ.* III. iv. (1873) 382.

Rheum officinale. [Extract.] *Pharm. Journ.* III. iv. (1874) 690–691.

Programme de Cours d'histoire naturelle médicale professé à la
faculté de médicine à Paris. 3ᵉ partie ; étude spéciale des
plantes employées en médicine. Paris, 1877. 18°.

The genus Copaifera. [*Journ. de Pharm.*] *Pharm. Journ.* III.
vii. (1877) 873–875.

BAKER (John Gilbert).

Note on Mikania Guaco. *Pharm. Journ.* III. xi. (1880) 471.

BALDAMUS (Ed.).

Die Erscheinungen der deutschen Literatur auf dem Gebiete der
Land-, Forst-, und Hauswirthschaft, sowie des Gartenbaues.
1871–1875. Alphabetisch geordnet und mit einem Materialen-
Register versehen. Leipzig, 1876. 8°.

BALFOUR (Edward).

Remarks on the Gutta Percha Tree of Southern India, noticing
also the history and manufacture of the Gutta-percha of
commerce. Madras, 1855. 8°. [*Anon.*]

The Cyclopaedia of India and of Eastern and Southern Asia,
commercial, industrial, and scientific, *etc.* Madras, 1857. 4°.
Ed. 2. 1871–73. 5 vols. 8°.

The Supplement to the Cyclopaedia of India and of Eastern and
Southern Asia, *etc.* Madras, 1858. 4°.

The Timber Trees, Timber, and Fancy Woods, as also the Forests
of India and of Eastern and Southern Asia. Ed. 2. Madras,
1862. 8°. Ed. 3. 1870.

BALFOUR (John Hutton).

Description of the plant which produces the Ordeal Bean of
Calabar. *Royal Soc. Edinb. Trans.* xxii. (1861) 305–312.

Remarks on plants furnishing varieties of Ipecacuan, and on the
cultivation of Cephaelis Ipecacuanha (Rich.) in the Royal
Botanic Garden of Edinburgh. *Pharm. Journ.* III. ii. (1872)
948–949, 969–971.

Rheum palmatum, var. tanguticum. [*Trans. Bot. Soc. Edinb.*]
Pharm. Journ. III. viii. (1878) 588.

Barbe (——) Étude sur l'olivier.

BALL (Samuel).

An Account of the Cultivation and Manufacture of Tea in China; derived from personal observation during an official residence in that country from 1804 to 1826; and illustrated by the best authorities, Chinese as well as European: with remarks on the experiments now making for the introduction of the culturè of the Tea Tree in other parts of the World. London, 1848. 8°.

BALMIS (Francisco Javier).

Demonstracion de las eficaces virtudes nuevamente descubiertas en las raices de los plantas de Nueva España, especies de Agave y de Begonia, *etc.* Madrid, 1794. 8°.

In German, Leipzig, 1797. In Italian, Roma, 1795.

BAMBER (E. F.).

Tea. London, 1868. 8°.

BANAL (Antoine).

Catalogue des plantes usuelles. Montpellier, [1755]. 8°.

Other editions dated, 1780, 1784, 1786.

BANCROFT (Joseph).

The newly introduced poisonous burr, Xanthium Strumarium. Brisbane, 1880. 8°.

BARBER (G.).

The pharmaceutical or medico-botanical map of the world, showing the habitats of all the medicinal plants and drugs in general use. London, 1869. fol.

BARBEU-DUBOURG (Jacques).

Le botaniste français, comprenant toutes les plantes communes et usuelles, *etc.* Paris, 1767. 2 vols. 8°.

BARHAM (Henry).

Hortus americanus: containing an account of the trees, shrubs, and other vegetable productions of South-America and the West-India Islands . . . their uses in medicine, diet and mechanics, *etc.* Kingston, Jamaica, 1794. 8°.

BARJAVEL (C. F. H.).

Traité complet de la culture de l'olivier, redigé d'après les observations et experiences de l'abbé F. Jamet. Marseille, 1831. 8°.

BARTLEY (George C. T.).

The Cultivation of Common Fruits, from an Economic and Social Point of View. *Journ. Soc. Arts*, xxv. (1877) 146–149.

BARTOLO (Giovanni di).

Della coltivazione del cotone secondo le antiche pratiche di Terranova in Sicilia. Torino, 1864. 16°.

BARTON (Benjamin H.) & Thomas CASTLE.

The British Flora medica. London, 1837–38. 2 vols. 8°.
[New edition, revised, condensed, and partly re-written by
J. R. Jackson] 1877.

BARTON (Benjamin Smith).

Collections for an Essay towards a Materia medica of the United
States. Philadelphia, 1798–1804. 8°.

BARTON (William P. C.).

Vegetable Materia medica of the United States; or medical botany;
containing a botanical, general and medical history of medicinal
plants indigenous to the United States. Philadelphia, 1817–18.
2 vols. 4°.

Some account of a plant used in Lancaster County, Pennsylvania,
as a substitute for chocolate (Holcus bicolor, Willd.). Philadelphia, 1816. 8°.

BARTOSSÁG (Joseph).

Gotterbaum (Ailanthus glandulosa). Ofen, 1841. 8°.

BARUCHSON (Arnold).

Beetroot Sugar : remarks upon the advantages derivable from its
growth and manufacture in the United Kingdom, *etc.* London,
1868. 8°. Ed. 2. 1870.

BASSET (Nicolas).

Traité pratique de la culture . . . de la betterave, *etc.* Paris,
1854. 12°. Ed. 2. 1858. Ed. 3. 1868.

Guide pratique de fabricant de sucre. Paris, 1861–65. 2 vols.
8°. Ed. 2. 1872–5. 3 vols. 8°.

La sucrerie indigène, étrangère et exotique. Paris, ——. 8°. *

Lettre à un raffineur sur la situation réelle de l'industrie sucrière
française. Paris, 1873. 8°.

BASTIDE (L.).

L'Alfa; végétation, exploitation, commerce, industrie, papeterie.
Oran, 1877. 8°.

BASROGER (J.).

Description des principaux champignons comestibles et des
champignons vénéneux avec lesquels ils peuvent facilement
confondus. Cluny, 1880. 16°.

BATCHELOR (H. W.).

Dika bread. [Extract.] *Pharm. Journ.* III. xi. (1880) 43.

BATEMAN (T. H.).

A few notes from the far East on Opium. *Pharm. Journ.* III. v.
(1875) 906–907.

BATES (G. Hubert).

 On the gathering and curing of Carrageen in Massachusetts. [U.S. Agricultural Report.] *Pharm. Journ.* II. xi. (1869) 298–302.

BATKA (Johann B.).

 Ueber Sarsaparill. Leipzig, 1834. 8°.

 On the plants from which Senna-leaves are obtained. *Pharm. Journ.* ix. (1849) 25–31.

 On Cinchona rubra and C. Savanilla. *Pharm. Journ.* xi. (1852) 321–323.

 Monographie der Cassiengruppe Senna. Prag, 1866. 4°.

BAUD (J. C.).

 Bijvoegsels tot de proeve eener geschiedenis van den handel en het verbruit van opium in Ned. Indië. *Neêrl. Ind. Bijdrag,* ii. (1854) 189–211.

BAUP (Samuel).

 On the resins of the " Arbol-a-brea and of Elemi." [*Journ. de Pharm.*] *Pharm. Journ.* xi. (1852) 313–318.

BAUR (B.).

 On Attar of Rose. [Transl. and abridged by D. Hanbury.] *Pharm. Journ.* II. ix. (1867) 286–293.

BAUR (Franz).

 Ueber forstliche Versuchsstationen. Ein Weck- und Mahnruf an aller Pfleger und Freunde des deutschen Waldes. Stuttgart, 1868. 8°.

 Der Wald und seine Bodendecke im Haushalte der Natur und der Voelker, *etc.* Stuttgart, 1869. 8°.

 Die Fichte in Bezug auf Ertrag, Zuwachs und Form, *etc.* Berlin, 1877. 8°.

BAYER (G. C.).

 Anleitung zum Anbau und zur Verwerthung der wichtigsten Handelsgewaechse. Hannover, 1838. 8°.

BAYLES (W. E.).

 Les produits commerciaux et industriels. I. Description, emploi, provenances et débouchés. Boulogne-sur-Mer, 1881. 8°.

BAZLEY (*Sir* Thomas).

 [A discussion opened by, on] The Claims of India Cotton. *Journ. Soc. Arts,* xix. (1871), 386–393.

BEAVER (S.).

 Fir-wool oil and Fir-wool. *Pharm. Journ.* II. iv. (1863) 424–425.

BEC (A. de).

 Culture du tabac en France, et particulièrement dans les Bouches-du-Rhône. Ed. 2. Aix, 1875. 16°.

BECHSTEIN (Johann Matthaeus).

Kurzgefasste gemeinnuetzige Naturgeschichte der Gewaechse des In- und Auslandes. Leipzig, 1796. 2 vols. 8vo.

Taschenblaetter der Forstbotanik. Die in Deutschland einheimischen und akklimatischen Baeume, Straeuche und Stauden enthaltend. Weimar, 1798. 8°. Ed. 2. by S. Behlen, 1828.

Forstbotanik, oder vollstaendige Naturgeschichte der deutschen Holzarten und einiger fremden. Erfurt, 1810. Ed. 5. by S. Behlen, 1843.

Forstbotanik. Zweiter Theil, enthaltend Forstkraueterkunde, oder Naturgeschichte der deutschen Forstkraeuter. Herausgegeben von Stephan Behlen und mitbearbeitet von F. A. Desberger. Erfurt und Gotha, 1833. 8°.

BECK (Otto).

Instruktion ueber das Pflanzen und die Pflege der Allerbaeume. Ed. 4. Trier, 1873. 8°.

Erster Jahresbericht [1872] ueber die Hebung des Flachs- und Hanfbaues im Reg. Bez. Trier. Trier, 1873. 8°.

Zweiter [1873] Jahresberichte ueber . . . im Regierungsbezirk Trier. Trier, 1874. 8°.

BECKFORD (William).

A descriptive Account of the Island of Jamaica, with remarks upon the Cultivation of the Sugar Cane, etc. London, 1790. 2 vols. 8°.

BECLU (H.).

Nouveau manuel de l'herboriste, ou traité des propriétés médicinales des plantes exotiques et indigènes du commerce, etc. Corbeil, 1872. 18°.

BEDDOME (Richard Henry).

The Trees of the Madras Presidency. Madras, 1863. 8°.

The Flora Sylvatica for Southern India, containing . . . plates of all the principal timber trees in Southern India and Ceylon, etc. Madras [1869–73]. 4°.

BEECHEY (Frederick William).

Notice of the Cloth Tree (Broussonetia papyrifera) of the South Sea Islands. [From Beechey's *Voyage to the Pacific*.] *Pharm. Journ.* x. (1850) 148.

BEICHE (W. Ed.).

Taschenbuch der Pflanzenkunde fuer Land- und Forstwirthe oder Beschreibung aller wichtigen Cultur-, Futtur-, und Unkrautpflanzens Deutschlands nebst Angabe ihres Nutzens und Schadens, etc. Berlin, 1869. 16°.

BEKE (Charles Tilstone).

On the Korarima, or Cardamom of Abessinia. *Pharm. Journ.* vi. (1847) 511–512.

BELANGER (Charles).

Essais de culture du Quinquina à la Martinique. Paris, 1870. 8°.

BELL (Jacob).

On the adulteration of Senna. *Pharm. Journ.* ii. (1842) 63–65.

East Indian Senna. *Pharm. Journ.* ix. (1850) 401–402.

BENTLEY (Robert).

Plant yielding the African Arrow-root. [Maranta arundinacea.] *Pharm. Journ.* x. (1850) 272.

On a species of Smilax, and a new commercial sort of Sarsaparilla which is obtained from it. *Pharm. Journ.* xii. (1853) 470–476.

On the characters of Dandelion root (Leontodon Taraxacum), and the means whereby it may be distinguished from other roots. *Pharm. Journ.* xvi. (1856) 304–307.

Culture of Medicinal Plants at Hitchin, Herts. *Pharm. Journ.* II. i. (1859) 275–279, 323–325, (1860) 414–416, 515–518.

Remarks on Taraxacum root. *Pharm. Journ.* II. i. (1860) 402–404.

Note on the Roman Chamomile (Anthemis nobilis). *Pharm. Journ.* II. i. (1860) 447–450.

A Manual of Botany. London, 1861. 8°. Ed. 2. 1870. Ed. 3. 1873.

On Actaea, or Cimicifuga racemosa. *Pharm. Journ.* II. ii. (1861) 460–468.

On the adulteration of Senna with Argel leaves. *Pharm. Journ.* II. ii. (1861) 496–498.

On the adulteration of Black Hellebore (Helleborus niger) with Bane-berry (Actaea spicata). *Pharm. Journ.* II. iii. (1861) 109–112.

Note on Feuillaea and other donations [to the Society's Museum. Abstract.] *Pharm. Journ.* II. iv. (1862) 198–199.

New American Remedies.—I. Podophyllum peltatum. *Pharm. Journ.* II. iii. (1862) 456–464.

———— [II.] Hydrastis canadensis, Linn.—Yellow root, Orange root. *Pharm. Journ.* iii. (1862) 540–546.

———— III. Xanthorrhiza apiifolia, Willd.—Yellow root. *Pharm. Journ.* iv. (1862) 12–14.

———— IV. Caulophyllum thalictroides, Michx.—Blue Cohosh. *Pharm. Journ.* iv. (1862) 52–56.

BENTLEY (Robert), *continued* :—

New American Remedies.—V. Jeffersonia diphylla, Pers.—Twin-leaf. *Pharm. Journ.* iv. (1862) 104–170.

—— VI. Sanguinaria canadensis, Linn.—Blood-root, Puccoon. *Pharm. Journ.* iv. (1862) 263–269.

—— VII. Sarracenia purpurea, Linn.—Indian Cup, Side-saddle flower. *Pharm. Journ.* iv. (1862) 294–302.

—— VIII. Dicentra (Corydalis) formosa, Borkh.—Turkey Corn. *Pharm. Journ.* iv. (1863) 353–357.

—— IX. Xanthoxylon fraxineum, Willd. —Prickly Ash, Toothache Shrub. *Pharm. Journ.* iv. (1863) 399–407.

—— X. Ptelia trifoliata, Linn.—Shrub Trefoil, Wafer Ash. *Pharm. Journ.* iv. (1863) 494–497.

—— XI. Geranium maculatum, Linn. —Spotted Cranesbill, Alum root. *Pharm. Journ.* II. v. (1863) 20–25.

—— XII. Cerasus (Prunus) virginiana, Michaux. — Wild Cherry, Wild Black Cherry. *Pharm. Journ.* v. (1863) 97–105.

—— XIII. Baptisia tinctoria, R. Brown.—Wild Indigo. *Pharm. Journ.* v. (1863) 211–216.

—— XIV. Hydrangea arborescens, Linn.—Common Hydrangea, Seven-barks. *Pharm. Journ.* v. (1864) 310–315.

On a new kind of Matico, with some remarks on officinal Matico. *Pharm. Journ.* II. v. (1864) 290–297.

On the organic Materia medica of the British Pharmacopoeia. *Pharm. Journ.* II. v. (1864) 416–428, 479–491.

On the adulteration of Saffron with the stamens of Crocus. *Pharm. Journ.* II. vii. (1866) 452–457.

On the study of botany in connection with pharmacy. *Pharm. Journ.* II. viii. (1866) 108–118.

Note on the substitution of Mangosteen (Garcinia Mangostana) for Bael (Aegle Marmelos). *Pharm. Journ.* II. viii. (1867) 654–655.

On the Study of Botany in connection with Pharmacy. [Address at Dundee.] *Pharm. Journ.* II. ix. (1867) 152–157.

On the characters, properties, and uses of Eucalyptus Globulus, and other species of Eucalyptus. London, 1874. 8°.

The Characters, properties, and uses of Eucalyptus Globulus and other species of Eucalyptus. [Abstract of a lecture.] *Pharm. Journ.* III. iv. (1874) 872–879.

On a drug substituted for Chiretta (Ophelia Chirata, Grisebach). *Pharm. Journ.* III. v. (1874) 481.

BENTLEY (Robert), *continued* :—

The admixture of White Hellebore with Valerian root. *Pharm. Journ.* III. vii. (1877) 649–650. *See also* p. 665.

Eucalyptus Globulus. *Pharm. Journ.* III. viii. (1878) 865–868.

BENTLEY (Robert), and Henry TRIMEN.

Medicinal plants, being descriptions, with original figures of the principal plants employed in medicine, and an account of the characters, properties, and uses of their parts and products of medicinal value. London, 18[75–]80. 4 vols. 8°.

BENNETT (Alfred William).

Notes on Indian Simarubeae. *Pharm. Journ.* III. iii. (1873) 801–802, 842–843, 881–883.

Notes on Indian Burseraceae. *Pharm. Journ.* III. vi. (1875) 62–64, 83–84, 102–104.

Additions and corrections to Notes on Indian Simarubeae. *Pharm. Journ.* III. vi. (1875) 162.

BENNETT (George).

On the introduction, cultivation, and economic uses of the Orange, and others of the Citron tribe, in New South Wales. [Sydney? 1867?] 8°.

The " Pituri," a new narcotic. [Extract.] *Pharm. Journ.* III. iv. (1873) 184.

BENNETT (John Whitchurch).

The Cocoa-nut tree, its uses and cultivation. London, [1827?]. 8°.

Ceylon and its capabilities : an account of its natural resources, indigenous productions and commercial facilities, *etc.* London, 1843. 4°.

BENNETT (John Joseph).

Description of a new species of Phrynium from Western Africa. [P. Danielli.] *Pharm. Journ.* xiv. (1854) 160–161.

Description of the Bungo, or Frankincense tree of Sierra Leone, regarded as a new genus of Caesalpiniae. [Daniellia thurifera, Bennett.] *Pharm. Journ.* xiv. (1854) 251–253.

Note on the species of Croton described by Linnaeus under the names of Clutia Eluteria and Clutia Cascarilla. [From the *Journal of the Linnean Society.*] *Pharm. Journ.* II. i. (1859) 132–134.

BENOIT (J. A.)

Culture speculative de la betterave au point de vue de l'industrie sucrière. Troyes, 1874. 8°.

2

Béraud (E.)
 Études sur le chêne et sur ses auxiliaires. Le Mans, 1864. 8°.
 Études forestières, No. 4. De la végétation spontanée des plantes
     naturelles forestières. Amiens, 1867. 8°.
Berchtold (Friedrich, *Grdf* von).
 Oekonomisch-technische Flora Boehmens, . . . Anwendung und
     Behandlung in Kuensten, Gewerben, Land-, Forst-, und
     Hauswirthschaft, *etc.* Prag, 1836–43. 3 vols. 8°.
        Unfinished ; ends with Pentandria,—Solanum.

Béranger (Adolfo di).
 Guida per il coltivatore di vivac boschivi, con cenni preliminari e
     note sulla materia forestale. Ed. 2. Firenze, 1880. 16°.
 Relazione sul pineto communale di Ravenna. 1880. 4°.
Berg (C. II. Edmund, *Freiherr* von).
 Geschichte der deutschen Waelder bis zum Schlusse des Mittel-
     alters. Ein Beitrag zur Culturgeschichte. Dresden, 1871. 8°.
Berg (Otto Carl).
 Handbuch der Pharmaceutischen Botanik fuer Pharmaceuten und
     Mediciner. Berlin, 1845. 8°. Ed. 3. 1858. Ed. 5. 1866.
 Characteristik der fuer die Arzneikunde und Tecknik wichtigsten
     Pflanzengenera, *etc.* Berlin, 1845. 4°. Ed. 2. 1861.
 Die Chinarinden der pharmacognostischen Sammlung zu Berlin.
     Berlin, 1865. 4°.
 Pharmaceutischer Waarenkunde. I. Pharmacognosie des Pflanzen-
     reichs. Ed. 3. Berlin, 1863. 8°. Ed. 4. by A. Garcke,
     1869.
 Anatomischer Atlas, zur pharmaceutischen Waarenkunde, *etc.*
     Berlin, [1863]. 4°. Ed. [2]. 1869.
Berg (Otto Carl), & C. F. Schmidt.
 Darstellung und Beschreibung saemmtlicher in der Pharmacopoea
     borussica aufgcfuehrten officinellen Gewaechse, oder der
     Theile und Rohrstoffe, welche von ihnen in Anwendung
     kommen. Leipzig, 1858–63. 4 vols. 4°. Ed. 2. [1863–69.]
Bergen (Heinrich von).
 Versuch enier Monographie der China. Hamburg, 1826. 4°.
Bergius (Peter Jonas).
 Materia medica, e regno vegetabili, sistens simplicia officinalia
     pariter atque culinaria. Stockholmiae, 1767. 2 vols. 8°. Ed.
     2. 1781.
Bergsma (Cornelis Adrian).
 De Thea. Trajecti ad Rhenum, 1824. 8°.

BERNARD (—.).

Mémoire pour servir à l'histoire naturelle de l'olivier. [Recueil
de l'academie de Montpellier.] Montpellier, 1783. 8°.

BERNARDIN (J.).

Les odeurs de Muse, de Vanille, et de Violette. [Annecy,
1870 ?] 8°.

Notice sur les collections scientifiques et sur le musée commercial-
industrielle. Gand, 1871. 8°. (Produits bruts, pp. 20–77;
Classification des huiles et des graisses végétales, pp. 37–47.)

Classification de 250 matieres tannantes. Gand, 1872. 8°.

Nomenclature usuelle 550 fibres textiles avec indication de leur
provenance, leurs usages, etc., par * * * Gand, 1872. 8°.
[*Anon.*]

Classification de 100 caoutchoucs et gutta-perchas suivie de notes
sur les sucs de Balata et de Massaranduba. Gand, 1872. 8°.
Supplement [1875].

Classification de 160 huiles et graisses végétales. 2ᵉ edition . .
Gand, 1874. 8°.

Classification de 40 savons végétaux. Gand, 1875. 8°.

Supplement à la Classification de 250 matieres tannantes. [Gand,
1875.] 8°.

Classification de 250 fécules. Gand, 1876. 8°.

BERNATZIC (M.).

A new falsification of Cinchona bark. [*Journ. de Pharm.*]
Pharm. Journ. III. iv. (1874) 569–570.

BERNAYS (Lewis A.).

The Olive and its Products. A treatise on the habits, cultivation, and propagation of the tree; and upon the manufacture of oil and other products therefrom. Brisbane, 1872.
8°.

BERNHARDT (August).

Ueber die historische Entwickelung der Waldwirthschaft und
Forstwissenschaft in Deutschland. Berlin, 1871. 8°.

BERTERO (Carlo Giuseppe).

Specimen medicum nonnullas indigenas stirpes continens exoticis
succedaneas. Taurini, 1811. 4°.

BERTHELOT (Marcellin).

On the Manna of Sinai, and the Manna of Syria. [From the
Comptes Rendus.] *Pharm. Journ.* II. iii. (1861) 274-275.

BESSE (A.).

Essais d'amélioration de la culture de la garance [Rubia sativa].
Avignon, 1877. 8°.

Besse (A.), & A. Rieu.

Essais d'amélioration de la culture de la garance. Mémoire présenté à la chambre de commerce d'Avignon et à la société d'agriculture de Vaucluse, le 20 Janvier, 1875. Avignon, 1875. 8°.

Bevan (E. J.), & C. F. Cross.

Contributions to the chemistry of bast fibres. Manchester, 1880. 8°.

Bialet (Dr. —.).

Maté, or Paraguayan tea. [Extract.] *Pharm. Journ.* III. vii. (1876) 4–5.

Bianchedi (C. A.).

L'olivicoltura e l'oleificazione nel circondario di Salo: studi e proposte. Brescia, 1877. 8°.

Bibra (Ernst, *Freiherr* von).

Die narkotischen Genussmittel und der Mensch. Nuernberg, 1855. 8°.

Bidie (George C. M.).

Report on the Ravages of the Borer in Coffee estates, with a review of the existing systems of coffee culture, and suggestions for the further development of the productive resources of the Coffee districts in Southern India. Madras, 1869. 8°.

Catalogue of the Raw Products of Southern India forwarded to the Paris Exhibition of 1878. Madras, 1878. 8°.

The Timber Trees of India. Madras, 1862. 8°.

Bigg (William).

Answers to queries respecting the cultivation of English Rhubarb near Banbury. *Pharm. Journ.* vi. (1846) 74–76.

Bigelow (J. M.).

American Eupatoria. [Extract.] *Pharm. Journ.* III. v. (1874) 303–304.

Billinger (Otto).

Historical notes on Opium. [Abstract.] *Pharm. Journ.* III. vii. (1876) 452–454, 1041–1042.

Birnbaum (K.).

Lehrbuch der Landwirthschaft. Frankfort-am-Main, 1859–63. 8°.

Birdwood (George C. M.).

Catalogue of the Economic products of the Presidency of Bombay; being a catalogue of the Government Central Museum. Division 1. Raw Produce (Vegetable). Bombay, 1862. 8°.

Catalogue of the vegetable productions of the Presidency of Bombay, including a list of the drugs sold in the bazaars of Western India. Second edition, with index. Bombay, 1865. 8°.

Frankincense, or Olibanum. [*Trans. Linn. Soc.*] *Pharm. Journ.* III. i. (1870) 163–167.

BIRKMEYER (Christian).

Populaere Pflanzenkunde und Hausapotheke, *etc.* Ed. 2. Neu-Ulm, 1868. 8°.

Kraeuterbuch nebst Hausapotheke. Ed. 3. Neu-Ulm, 1871. 8°.

BISCHOFF (G.).

On the botanical origin of Senna leaves. [From the *Central Blatt.* See also *Pharm. Journ.* ix. 25.] *Pharm. Journ.* x. (1851) 543–546.

BISCHOFF (Gottlieb Wilhelm).

Plantae medicinales . . . adjectis medicamentis, quae praebent, simplicibus. Heidelbergae, 1829. 4°.

Grundriss der medicinischen Botanik als Leitfaden bei Vorlesungen, so wie zum Selbststudium, und besonders zum repetitorischen Studium fuer Studirende, *etc.* Heidelberg, 1831. 8°.

Medicinisch-pharmaceutische Botanik. Ein Handbuch fuer Deutschlands Aerzte und Pharmaceuten. Erlangen, 1843. 8°. Ed. 2. 1847.

BIUSO (S.)

Monografia sul fico d'India in Sicilia. Palermo, 1879. 8°.

BIZIO (G.)

Il caffé. Venezia, 1870. 8°.

BLACKWELL (Elizabeth).

A curious herbal, containing 500 cuts of the most useful plants, which are now used in the practice of physick, *etc.* London, 1737[-39]. 2 vols. fol.

Herbarium Blackwellianum emendatum et auctum, . . . Edidit . . . C. J. Trew, *etc.* Nuernberg, 1750–73. 3 vols. fol.

BLAGRAVE (Joseph).

Supplement or enlargement of Nich. Culpeper's English Physician, *etc.* London, 1666. 8°. Ed. 2. 1674.

BLANCHET (C.)

Thapsia garganica, or Bou-Nefa of the Arabs. [Abstract.] *Pharm. Journ.* III. x. (1880) 889–891.

BLANKAART (Stephan).

De Nederlandschen Herbarius, or kruidboek der vornaamste kruiden, tot de medicyne, spysbereidingen en konstwerken dienstig, handelende van zommige hier te lande wassende boomen, kruiden, heesters, mossen, enz. Amsterdam, 1698. 8°. Ed. 2. 1714.

BLASKOVICS (E. von).

Die Sojabohne [Soja hispida]. Etwas ueber deren Cultur, Verwendbarkeit und Werth als Futtermaterial. Wien, 1880. 8°.

BLASQUEZ (Pedro), & Ignazio BLASQUEZ.

Memorie sobre el Maguey mexicano. (Agave Maximilianea.) Mexico, 1865. 8°.

BLEEKRODE (S.).

The Chinese green colour Lo-Kao, and some other dyes. *Pharm. Journ.* II. i. (1859) 228–234.

The Soap-bark of South America. [From *Journ. Soc. Arts.*] *Pharm. Journ.* II. i. (1860) 471–473.

BLIJHAM (G.)

Schetsen van in Nederland voorkommende vergiftige planten, *etc.* Amsterdam, 1880. 8°.

BLONDEL (J.).

Manuel de la fabrication du sucre de betteraves. [Peronne,] 1863. 8°. Ed. 2. Douai, 1870.

BLYTH (A. Wynter).

Piper nigrum. [*Chem. News.*] *Pharm. Journ.* III. vi. (1875) 303–305.

> The last page has a bibliography of the subject.

BODARD (Pierre Henri Hippolyte).

Analyse du cours de botanique médicale-comparée, *etc.* Paris, 1809. 4°.

Cours de botanique médicale comparée, ou exposé des substances végétales exotiques comparées aux plantes indigènes. Paris, 1810. 2 vols. 8°.

BODIN (Jean).

Herbier agricole, ou liste des plantes les plus communes, à l'usage des écoles d'agriculture, *etc.* Paris, 1856. 18°. [Ed. 2. 1862.] 8°. Ed. 6. Corbeil, 1880. 18°.

BOEHMER (Georg Rudolf).

Technischen Geschichte der Pflanzen welche bei Handwerken, Kuensten und Manufacturen bereits in Gebrauche sind, odor noch gebraucht werden koennen. Leipzig, 1794. 2 vols. 8°.

BOETTGER (Rudolph).

A simple method of detecting fibres of cotton in linen textures. [From the *Annalen der Chemie.*] *Pharm. Journ.* iii. (1844), 351.

BOITEL (A.).

Du pin maritime, de sa culture dans les dunes, de la pratique du résinage et de l'industrie des résins, *etc.* Paris, 1848. 8°.

Culture du cédratier en Corse. Paris, 1875. 8°.

BOLAS (Thomas).

The India-Rubber and Gutta-Percha Industries. *Journ. Soc. Arts*, xxviii. (1880) 753–760, 763–767, 773–779, 783–789, 793–796, 803–808. Also published separately as Cantor Lectures.

BONAFOUS (Mathieu).

De la culture des muriers. Lyon, 1822. 8°. Ed. 2. 1824. Ed. 3. Paris, 1827. Ed. 4. 1840.

Histoire naturelle, agricole et économique du Mays. Paris, 1836. fol.

BONAVIA (E.).

Notes on the medicinal properties of the Thistle-oil, or oil extracted from the seed of the Argemone mexicana. [Extract.] *Pharm. Journ.* II. viii. (1866) 297–299.

BONNET (Henri).

La truffe. Études sur les truffes comestibles au point de vue botanique, entomologique, forestier et commercial. Paris, 1869. 8°.

BONTEKOE (Cornelis).

Tractat van het excellenste kruyd *thee.* Benevens een kort discours op het leven, de siekte en de dood. Den tweeden druk vermeerdert . . . van twee . . . verhandelingen. I. Van de coffi. II. Van de chocolate. 's Gravenhage, 1679. 8°. Ed. 3. 1685.

BONYAN (George R.).

On Spigelia Anthelmia, or Demerara Pink-root. *Pharm. Journ.* v. (1846) 354, 355.

BONZOM (E.), —. DELAMOTTE, & Ch. RIVIÈRE.

Du caroubier et de caroube. Plantation et greffage du caroubier en Algérie. Nourriture des animaux domestiques par les caroubes. Essai de propagation. Paris, 1878. 8°.

BOOTH (John).

Die Douglas-Fichte und einige andere Nadelhoelzer namentlich aus dem nordwestlichen Amerika in Bezug auf ihren forstlichen Anbau in Deutschland. Berlin, 1877. 8°.

BOOTH (Francis).

Two introductory lectures on (vegetable) Materia medica. London, 1827. 8°.

BORDIGA (Oreste), & Leopoldo SILVESTRINI.

Del riso e della sua coltivazione : studia di economia rurale del prof. O. B., e di chimica agraria del prof. L. S.; con un' aggiunta d'uno studio sulle questioni economiche ed igieniche riflettenti il commercio e la coltivazione del riso, del prof. O. B., *etc.* Novara, 1880. 8°.

BORGGREVE (Bernard).

Haide and Wald. Spezielle Studien und generelle Folgerungen ueber Bildung und Erhaltung der sogenannten natuerlich. Vegetationsformen oder Pflanzengemeinen. Fuer Botaniker, Geographien, Staats-, Land-, und Forstwirthe. Berlin, 1875. 8°.

BORKHAUSEN (Moritz Balthasar).

Theoretisches praktisches Handbuch der Forstbotanik und Forsttechnologie. Giessen und Darmstadt, 1800–3. 2 vols. 8°.

BORN (L.)

Der Mais als Futtermittel. Jena, 1880. 8°.

BOSCH (Miguel).

Manual de botanica applicada á la agricultura y á la industria. Madrid, 1858. 8°.

BORSÌ (A.).

Flora forestale italiana, ossia descrizione delle piante legnose indigene all' Italia. Firenze, 1880. 8°.

BOSISTO (—.).

Is the Eucalyptus a fever-destroying tree? [*Melbourne Argus.*] *Pharm. Journ.* III. v. (1874) 270.

Ueber Eucalyptus und ihre Eigenschaften. Aus dem Englischen von Antoine. *Oesterr. bot. Zeitschr.* xxx. (1880) 20–23.

BOSSU (Antonin).

Traité des plantes médicinale indigènes. Paris, 1854. Ed. 2. 1862. Ed. 3. 1872.

BOUASSE-LEBEL.

Encyclopédie Bouasse-Lebel. Histoire naturelle; Plantes utiles. [2 parts.] Paris, 1876. 18°.

BOUCHARDAT (Apollinaire).

On the Vanilla of the Island of Bourbon. [From the *Journ. de Pharm.*] *Pharm. Journ.* ix. (1849) 275–276.

Adulteration of Pepper. [Extract.] *Pharm. Journ.* III. iii. (1873) 993–994.

BOUCHÉ (C. B.), & Hermann GROTHE.

Die Nessel als Textilpflanze. Berlin, 1877. 8°.

BOULIER (Ch.).

Collection of Scammony in the North-West of Asia Minor. [Extract.] *Pharm. Journ.* II. i. (1860) 521–523.

BOUQUET DE LA GRYE (Amédée).

Guide du forestier: 1ʳᵉ partie; Elements de sylviculture. 2ᵉ partie; Surveillance des foréts. Ed. 6. Paris, 1872. 3 vols. 18°.

BOUQUET DE LA GRYE (Amédée), *continued :*—
  Les bois indigènes, *etc.* See DUPONT (A. E.), & BOUQUET DE LA
  GRYE.

BOUQUINAT (F.).
  Traité sur les arbres résineux. Culture et produits. Châlons-
  sur-Saône, 1875. 12°.

BOURGOIN D'ORLI (P. H. F.).
  Guide pratique de la culture du caféier et du cacaoyer, suivi de la
  fabrication du chocolat. Paris, 1867. 12°.
  Guide pratique de la culture de la canne à sucre. Paris, [no date.]
  12°.

BOUTCHER (William).
  Treatise on Forest Trees. Edinburgh, 1775. 4°. Ed. 2. 1778.

BOUTON (Louis S.).
  Medicinal plants growing or cultivated in the Island of Mauritius.
  Mauritius, 1857. 8°.
  Plantes médicinales de Maurice. Ed. 2 [of the last]. Port-Louis,
  1864. 8°.
  Rapport présenté à la chambre d'agriculture, sur les diverses
  espèces de cannes a sucre cultivées a Maurice. [Port Louis,]
  Maurice, 1863. 8°.

BRADLEY (Richard).
  The Virtue and Use of Coffee with regard to the Plague, and
  other Infectious Distempers; containing the most Remarkable
  Observations of the Greatest Men in Europe concerning it, from
  the first Knowledge of it, down to the Present Time. To which
  is prefixed an exact figure of the Tree, Flower, and Fruit,
  taken from the Life. London, 1721. 8°.
  The Riches of a Hop-Garden explain'd. From several Improvements arising by that Beneficial Plant. As well to the private
  cultivators of it, as to the publick. London, 1729. 8°.

BRADY (Henry Bowman).
  On the Anatomy of Drugs. *Pharm. Journ.* II. viii. (1867) 408–
  412, 455–463, 544–550.
  Note on Hungarian Red Pepper. *Pharm. Journ.* III. xi. (1880)
  469.

BRAGGE (William).
  Bibliotheca nicotiana; a catalogue of books about Tobacco, *etc.*
  [Birmingham], 1880. 4°.

BRANDIS (Dietrich).
  Report on the Pegu Teak forests for the years 1857–60, *etc.*
  Calcutta, 1861. 8°.

BRANDIS (Dietrich), *continued :*—

List of specimens of some of the woods of British Burmah sent to England for the Exhibition of 1862. Rangoon, 1862. 4°.

Supplement to Reports on Forest Management in France, Switzerland, and Lower Austria. [London, 1874.] 8°.

Catalogue of specimens of Timber, Bamboos, Canes, and other Forest Produce from the Government Forests in the Provinces under the Government of India and the Presidencies of Madras and Bombay,-sent to the Paris Exhibition of 1878, by Order of the Government of India. Calcutta, 1878. 8°.

Forest Flora. *See* STEWART (J. L.), & D. BRANDIS.

BRANDT (Johann Friedrich von).

Tabellarische Uebersicht der officinellen Gewaechse. Berlin, 1829–30. fol.

BRANSON (W. P.).

Coffee : A Review of the Present Condition of its Growth, with a Consideration of its Treatment and Consumption in the United Kingdom. *Journ. Soc. Arts,* xxii. (1874) 456–461.

BRAUN (E.) Der sogenannte rationelle Waldwirth, insbesondere die Lehre von der Abkürzung des Umtriebs der Waelder, *etc.* Frankfurt-am-Main, 1865. 8°.

BRAUNGART (R.).

Die Cultur, Statistik und Handelsverhaeltnisse des Hopfens in England. *Zeitschr. fuer die ges. Brauwesen,* iii. (1880) p. 11. *See also* C. WHITEHEAD.

BRAY (J. de).

La ramie . . . 2ᵉ édition, considérablement augmentée. Ouvrage à l'usage des colons et des écoles primaires rurales de l'Algérie. Paris, 1879. 12°.

BREITENLOHNER (Jakob).

Die Cultur der Korbweide. Prag, 1877. 8°.

BREFFAUT.

La saliciculture et la vannerie à Bussières-les-Belmont. Langres, 1874. 16°.

BRIERRE DE BOISMONT (Alexandre Jacques François), & —. POTTIER.

Élémens de botanique, ou histoire des plantes considerées sous le rapport de leurs propriétés médicales et de leurs usages dans l'économie domestique et les arts industriels. Paris, 1825. 12°.

BRIOSI (Giovanni).

Coltivazione sperimentale di sementi di tabacchi esteri e di piante foraggire, raccomandata per paesi meridionali. Roma, 1879. 4°.

BRITTEN (James).

The odours of plants. [*Gardeners' Chron.*] *Pharm. Journ.* III.
ii. (1872) 546.

BRODIGAN (Thomas).

A Botanical, Historical, and Practical Treatise on the Tobacco
Plant, in which the Art of Growing and Curing Tobacco in
the British Isles is made familiar to every capacity. London,
1830. 8°. *

BROILLARD (Charles).

Le traitement des bois en France, à l'usage des particuliers.
Nancy, 1881. 8°.

BROTHERTON (W.).

The Olive, The Rape, and The Flax: their products, and their
cultivation. London, 1857. 8°.

BROUGHTON (J.).

First report on Cinchona grown on the Neilgherry hills. [*Madras
Quarterly Journ.*] *Pharm. Journ.* II. ix. (1867) 239–243.

On a false cinchona bark of India. [Hymenodictyon excelsum,
Wall.] *Pharm. Journ.* II. ix. (1868) 418–421.

Hybridization of Cinchonae. [*Linn. Soc.*] *Pharm. Journ.* III.
i. (1870) 118.

Cinchona cultivation in India. *Pharm. Journ.* III. ii. (1872)
705–707.

Chemical examination of the bark of the Azadirachta indica.
[*Madras Journ. Med. Science.*] *Pharm. Journ.* III. iii. (1873)
992.

BROUSSE (P.).

Quelques mots sur l'étude des fruits. Montpellier, 1881. 4°.

BROWN (J. C.).

Reboisement in France; or records of the replanting of the Alps.
London, 1876. 8°.

Pine plantations on the sand wastes of France. Edinburgh,
1878. 8°.

BROWN (James).

The Forester, being plain and practical directions on the planting,
rearing, and general management of forest trees. Edinburgh,
1847. 8°. Ed. 2. Edinburgh and London, 1851. Ed. 3. 1861.
Ed. 4. Edinburgh, 1871.

BROWN (Robert), *of Campster*.

On the vegetable products used by the North-West American
Indians as food and medicine, in the arts, and in super-
stitious rites. *Pharm. Journ.* II. x. (1868) 89–94, 168–174.

Browne (Daniel J.)

The Sylva americana; or a description of the forest trees
indigenous to the United States, *etc.* Boston, 1832. 8°.

Brownen (George).

Seeds in drug-parcels; experiments on their vitality and con-
sequent usefulness. *Pharm. Journ.* III. ii. (1871) 201–202.

Bruyn Kops (G. F. de). *See* De Bruyn Kops.

Bryant (Charles).

Flora diaetetica, or history of esculent plants both domestic and
foreign, *etc.* London, 1783. 8°.

Buchenau (Franz).

Die botanischer Produkte der Londoner internationalen Industrie-
Ausstellung. Ein Bericht. Bremen, 1863. 8°.

Buc'hoz (Pierre Joseph).

Dictionnaire raisonné et universel des plantes, arbres et arbustes
de la France . . . considérés relativement à l'agriculture,
au jardinage, aux arts et métiers, à l'economie domestique
et champétre, et à la medicine des hommes et des animaux.
Paris, 1770. 4 vols. 8°.

Manuel médical et usuel des plantes tant exotiques qu'indigènes.
Paris, 1770. 2 vols. 8°.

Manuel alimentaire des plantes tant indigènes qu'exotiques qui
peuvent servir de nourriture et de boisson aux differens
peuples de la terre. Paris, 1771. 8°.

* Dissertation sur l'utilité et les bons et mauvaises effets du tabac,
du café, du cacao, et du thé. Paris, 1775. 8°. Ed. 2. 1788.

Herbier colorié des plantes mèdicinales de la Chine. Paris,
1781–91. fol.

Médicine végétale, tirée uniquement des plantes usuelles, *etc.*
Ed. 2. Paris, 1784. 8°.

Dissertation sur le café, sa culture, ses differentes préparations,
etc. Paris, 1785. fol. Paris et Liége, 1787. 8°.

Dissertation sur le putiet [Cerasus Padus], et ses propriétés recem-
ment découvertes; sur le laurier-amandier [C. Lauro-Cerasus] et
ses qualités délétères, et sur le bois de Sainte-Lucie [Prunus
Mahaleb] et ses propriétés économiques, *etc.* Paris, 1785. fol.

Dissertation sur le quassi (Cassis [Ribes nigrum]) et sur ses propriétés médicinales nouvellement découvertes. Paris, 1785. fol.

Dissertation sur le Quinquina, *etc.* Paris, 1785. fol.

Dissertation sur le rocoulier [Bixa Orellana] sur sa culture en
Europe et en Amérique . . . pour la teinture et pour d'autres
usages économique. Paris, 1785. fol.

Buc'hoz (Pierre Joseph), *continued.*:—

Dissertation sur l'Ipo [Upas], espéce de poison subtil, dont se servent les sauvages pour empoissoner les flèches. Paris, 1785. fol.

Etrennes du printemps aux habitants de la campagne et aux herboristes, ou pharmacie champétre, végétale et indigéne, à l'usage des pauvres et des habitants de la campagne. Ed. 5. Londres et Paris, 1785. 8°.

Dissertation sur le cachou, sur l'arbre d'ou on le tire, et sur ses propriétés pour la peinture. Paris, 1786. fol.

Dissertation sur le durion . . . qui merite d'être cultivé dans nos colonies. Paris, 1788. fol.

Dissertation en forme de supplément sur les plantes qui peuvent remplacer le thé. Paris, 1786. fol.

Dissertation sur les roses, leurs propriétés medicinales et économiques, *etc.* Paris, 1786. fol.

Dissertation sur la Brucée . . . comme specifique contre la dyssenterie. Paris, 1787. fol.

Dissertation sur la betterave . . . et la methode pour en tirer du sucre propre à remplacer le vrai sucre. Paris, 1787. fol.

Dissertation sur le cacao et sur sa culture. Paris et Liége, 1787. 8°.

Dissertation sur la chausse-trappe [Centranthus Calcitrapa], reconnue comme specifique dans les fièvres intermittentes, propre à remplacer le quinquina. Paris, 1787. fol.

Dissertation sur le blé du Turquie . . . et comme fourrage pour les bestiaux. Paris, 1787. fol.

Dissertation sur le Mangostan, un des arbres plus utiles de l'Inde, tant comme aliment que comme médicament, digne d'être transporté dans nos colonies de l'Amerique. Paris, 1787. fol.

Dissertation sur l'Illecebra, ou petite joubarbe . . . comme specifique contre le cancer, le charbon et la gangrène. Paris, 1787. fol.

Dissertation sur l'anis etoilé, *etc.* Paris, 1788. fol.

Presents de Flore à la nation française, pour les aliments, les médicaments, l'art vétérinaire et les arts et métiers. Ed. 2. Paris, 1787. 2 vols. 4°.

Dissertation sur le tabac et sur ses bons et mauvais effets. Paris et Liège, 1787. 8°. Ed. 2. (De la culture du tabac, *etc.*) 1800.

Dissertation sur le thé, sur sa récolte et sur les bons et mauvais effets de son infusion. Paris, 1786. fol.

Histoire universelle du règne végétal ou nouveau dictionnaire physique et économique de toutes les plantes qui croissent sur la surface du globe, *etc.* Paris, 1775–78. 12 vols. fol.

Unfinished, ends with Penn.

Buc'hoz (Pierre Joseph), *continued :*—

Dissertation sur l'ortie grièche. Paris, 1798. fol.

Manuel cosmétique et odoriférant des plantes. Paris, 1799. 8°.

Manuel économique des plantes, faisant suite aux manuels vétérinaire et tinctorial des plantes. Paris, 1799. 8°.

Manuel tinctorial des plantes, ou traité de toutes les plantes qui peuvent servir à la teinture et à la peinture. Ed. 5. Paris, 1800. 8°.

Traité ou manuel vétérinaire des plantes qui peuvent servir de nourriture et des médicaments aux animals domestiques. Ed. 2. Paris, 1801. 8°.

Le laurier et l'olivier réunis, entrelacés et considérés sous tous les aspects possibles, ou histoire naturelle du laurier et de l'olivier. Paris, An. x. [1802]. 8°.

Avantage qu'on peut tirer des plantes . . . pour guérir les maladies, *etc.* Paris, 1806. 8°.

Histoire naturelle du thé de la Chine . . . à laquelle on a joint un mémoire sur le thé du Paraguay, de Labrador, des Isles, du Cap, du Mexique, d'Oswégo, de la Martinique, etc., suivi d'une notice sur le cachou, le ginseng et l'huile de Cageput. Paris, 1806. 8°.

Burckhardt (Heinrich).

Ueber Eichenzucht. [Abdruck aus dem *Hannoverschen Land- und Forstwissenschaftlichen Vereinsblatte,* 1862.] Hildesheim, 1862. 8°.

Saeen und Pflanzen nach forstlicher Praxis. Handbuch der Holzerziehung. Forstwirthen, Forstbesitzern und Freunden des Waldes gewidmet. Ed. 5. Hannover, 1880. 8°.

Buhse (F. A.).

On the Gum Ammoniacum plant. [From the *Central Blatt.*] *Pharm. Journ.* xi. (1852) 526.

On the Galbanum plant. [From the *Central Blatt.*] *Pharm. Journ.* xi. (1852) 577.

Buisson (L.).

Étude sur le Cundurango de Loja. Précédée d'une notice historique et botanique. Paris, 1872. 8°.

Buist (George).

On some of the Undeveloped Resources of India. *Journ. Soc. Arts,* II. (1854) 315.

Bull (B. W.)

Remarks upon the Cinchona Pitaya, or Pitaya Bark. [From the *N. Y. Reg. of Med.*] *Pharm. Journ.* xi. (1851) 168–169.

BURBIDGE (F. W.).

The Gardens of the Sun, or a naturalist's journal . . . in . . .
Borneo and the Sulu archipelago. London, 1880. 8°.
Contains scattered notes on local economic botany.

BUREAU (Édouard).

De la famille de Loganiacées et des plantes qu'elle fournit à la
médicine. Thése. Paris, 1856. 4°.

BURGER (A.).

Du chéne du marine. Aperçu sur la production actuelle et
future de nos foréts domaniales. Paris, 1864. 8°.

BURGERSTEIN (A.)

Ueber die wichtigsten Gespinnstpflanzen. *Schriften naturw.
Kenntw. in Wien*, xix. (1880) 245.

BURGH (Nicholas Procter).

The Manufacture of Sugar, and the machinery employed for
Colonial and Home purposes. London, [1866]. 8°.

BURGSDORFF (Friedrich August Ludwig von).

Versuch einer vollstaendigen Geschichte vorzueglicher Holzarten in
systematischen Abhandlungen zur Erweiterung der Naturkunde
und Forsthaushaltungswissenschaft. Berlin, 1783–87. 2 vols. 4°.

Einleitung in die Dendrologie oder systematischer Grundriss der
Forstnaturkunde und Naturgeschichte. Berlin, 1800. fol.
Ed. 4. 1812.

—— [Transl. by J. J. Baudrillard] Nouveau manuel forestier.
Paris, 1808. 2 vols. 8°.

BURNETT (M. A.).

Plantae Utiliores; or, Illustrations of Useful Plants, employed in
the Arts and Medicine. London, 1842. 4 vols. 4°.
Without pagination; each part has a separate title-page.

BURNICHON (—.).

Notice sur la fleur du pyrèthere, dite insecticide Burnichon, *etc.*
Paris, 1863. 32°.

BURNOUF (Émile).

L'indigo japonais. Culture et préparation. Notice traduite pour
la première fois du japonais. Paris, 1874. 8°.

BURRELL (A.)

Indian Tea Cultivation: its Origin, Progress, and Prospects.
Journ. Soc. Arts, xxv. (1877) 199–215.

BUVRY (L.).

Anbauversuche mit auslaendischen Nutzpflanzen in Deutschland,
angestellt auf Veranlassung der Akklimatisations-Vereins in
Berlin. Berlin, 1868. 8°.

BYASSON (H.).

Note on Maté or Paraguay tea. [*Repert. de Pharm.*] *Pharm. Journ.* III. viii. (1878) 605–606.

CABANIS (F.).

Le mûrier, ses avantages et son utilité dans l'industrie. Paris, 1866. 8°.

CAILUS, DE. *See* DE CAILUS.

CALAU (Franz).

On Rhubarb. [From *Gauger's Repert.*] *Pharm. Journ.* ii. (1843) 658–660.

On Rad. Ginseng. From a Chinese communication. [From *Gauger's Repert.*] *Pharm. Journ.* ii. (1843) 661–662.

CALDAS (J. J.).

Description of the genuine Quina-tree of Loxa (Cinchona officinalis, now called Condaminea). [From the Regensburg *Flora.*] *Pharm. Journ.* xii. (1852) 20–22.

CALDER (James Erskine).

On the forests of Tasmania. *Proc. Royal Colonial Institute,* iv. (1873) 173–181.

The Woodlands, etc., of Tasmania. *Proc. Royal Colonial Institute,* v. (1874) 166–178.

CALKINS (Alonzo).

Opium and the Opium-appetite: with notices of alcoholic beverages, Cannabis indica, Tobacco and Coca, and Tea and Coffee, in their hygienic aspects and pathologic relations. New York, 1871. 8°.

CALVEL (Étienne).

De la betterave et de sa culture, considerée sous le rapport du sucre qu'elle renferme et particulièrement de la betterave de Castelnaudary. Paris, 1811. 8°.

CALVERT (F. Crace). *See* CRACE-CALVERT.

CALWER (C. G.).

Landwirthschaftliche und technische Pflanzenkunde Deutschlands. Stuttgart, 1852–55. 4°.

CAMERON (—.).

Java Cinchona bark. [Extract from Consular Report.] *Pharm. Journ.* III. xi. (1880) 52.

CAMINHOÁ (J. M.).

Das plantas toxicas do Brazil. Rio de Janeiro, 1871. 8°.

———— Catalogue des plantes toxiques du Brésil. Traduit du portugais par [F.] Rey. Paris, 1880. 8°.

CAMPBELL (Archibald).

On Indian Teas, and the Importance of Extending their Adoption in the Home Market. *Journ. Soc. Arts*, xxii. (1873) 173–181.

CANDOLLE (Casimir de).

Production of Cork. [Extract.] *Pharm. Journ.* II. iv. (1862) 89.

CANSTEIN (Philipp, *Baron* von).

Karte von der Verbreitung der nutzbarsten Pflanzen ueber der Erdkoerper. Berlin, 1834. fol. Text in 8°.

CANTONI (G.).

Études pratiques sur la culture du lin. [Extract.] Paris, 1876. 8°.

L'industria del tabacco. Parte I. La produzione. (*Annali di agricoltura*, Roma, 1879, No. 19.)

CAPPI (Giulio).

Non piu mancanza di foraggio, ossia la Penicellaria spicata. Piacenza, 1864. 8°.

La coltivazione dell' olivo e l'estrazione dell' olio nelle provincie oleifere italiane poste a confronto con i metodi piu razionali e moderni, opera originale illustrata con disegni delle macchine piu perfette. San Remo, 1875. 8°.

CARCENAC (Henry).

Le coton et sa culture; les plantes textiles. Paris, 1867. 8°.

Des textiles végétaux et des laines en Italic, en Espagne et en Portugal. Paris, 1869. 8°.

CARL (Joseph Anton).

Botanisch-medizinischer Garten, worinnen die Kraeuter in nahrhafte, heilsame und Giftige eingetheilt sind. Ingolstadt, 1770. 8°.

CARLES (P.).

The distribution of alkaloids in Cinchona barks. [Extract.] *Pharm. Journ.* III. iii. (1873) 643–644.

A new variety of Opium. [Extract.] *Pharm. Journ.* III. iii. (1873) 883.

CARLOTTI (Reg.).

La ramie, plante textile. Importance et utilité da sa culture en Corse. Ajaccio, 1877. 8°.

CARNAC (Harry Rivett).

Report on the Cotton department for the year 1868–69. Bombay, 1869. 8°.

CARPI (Am.).

L'Eucalyptus Globulus dal punto di vista igienico e terapeutico: memoria. Milano, 1879. 16°.

CARRÉ (Charles).

Étude sur les arbres à fruits à cidre dans le département de la
Sarthe. Mes loisirs en 1873. Paris, 1874. 4°.

CARUSO (Giralamo).

Trattato sulla coltivazione degli olivi e la manifattura dell' olio.
Palermo, 1870. 8°.

CASAUX (C. de).

Essai sur l'art de cultiver la canne et d'en extraire le sucre.
Paris, 1781. 8°.

> Published under the initials of C . . . x.

CASSELS (Walter R.).

Cotton: an account of its Culture in the Bombay Presidency,
prepared from Government Records and other authentic
sources, in accordance with a resolution of the Government
of India. Bombay, 1862. 8°.

CATTANEO (A.).

I miceti degli agrumi. Milano, 1879. 8°.

CATTANEO (Gottardo).

Della riacclimatazione del gelso. Milano, 1865. 8°.

CAUVET (D.).

The distinctive characters of Rhubarbs. [*Journ. de Pharm.*]
Pharm. Journ. III. ii. (1872) 1009–1010.

CAZAUD (—.).

Account of a new method of cultivating the Sugar Cane. London,
1779. 4°. (In English and French.)

CAZENEUVE (P.).

Microchemical examination of Angustura bark. [Abstract.]
Pharm. Journ. III. v. (1874) 7–8.

CAZIN (F. J.).

Traité pratique et raisonné des plantes médicinales indigènes et
acclimatées, avec un atlas de 200 plantes lithographiées.
4ᵉ édition . . . par H. Cazin. Paris, 1875. 8°.

CAZZUOLA (Ferdinando).

Il regno vegetale tessile e tintoriale ovvero descrizione delle
piante indigene ed esotiche che somministrano materie filabile
e coloranti. Firenze, 1875. 8°.

Dizionario di botanica applicata alla medicina, alla farmacia, alla
veterinaria, all' orticultura, all' agricoltura, all' industria e el
commercio. Pisa, 1879. 16°.

Le piante utile e nocive che crescono spontanee e coltivate in
Italia; con brevi cenni sopra la coltura, sopra i prodotti e
sugli usi che se ne fanno; ad uso di tutte le scuole del regno
d'Italia, *etc.* Torino, 1880. 8°.

CERFBERR DE MÉDELSHEIM (Alphonse E.).

Le cacao et le chocolat considerés aux points de vue hygièniqne, agricole, et commerciale. Paris, 1867. 18°. Ed. 2. [1873]. 8°.

CHABAUL (B.).

Végétaux exotiques cultivés en plein air dans la region des orangers (France). Toulon, 1872. 8°.

CHALONER (Edward), and —. FLEMING.

The Mahogany Tree : its botanical characters, qualities and uses, with practical suggestions for selecting and cutting it in the regions of its growth, in the West Indies and Central America, with notices of the projected interoceanic communication of Panama, Nicaragua, and Tehuantepec, in relation to their productions, and the supply of fine timber for ship-building and all other purposes. Liverpool, [1851]. 8°.

CHAMISSO (Adalbert von), *formerly* Louis Charles Adelaide CHAMISSO DE BONCOURT.

Uebersicht der nutzbarsten und schaedlichsten Gewaechse, welche wild oder angebaut in Norddeutschland vorkommen, *etc.* Berlin, 1827. 8°.

CHAMPION (P.), & H. PELLET.

De la betterave à sucre. Generalités sur la culture, influence de la graine, de l'ecartement, des engrais, etc. Paris, [1876]. 8°.

CHAPMAN (John).

The Cotton and Commerce of India, considered in relation to the interests of Great Britain. London, 1851. 8°.

CHAPPELLIER (P.).

Note sur le safran. (Bull. Soc. imp. d'acclimatation, Mai, 1862.) Note sur l'origine du Crocus sativus (Safran officinal). Paris, [1873]. 8°.

De la culture des safrans étrangers introduits en France par la société d'acclimatation. Paris, [1874]. 8°.

CHARLEY (William).

Flax and its products in Ireland. London, 1862. 8°.

CHARRIER (—.).

L'alfa des hauts plateaux de l'Algérie. Alger. 1873. 8°.

CHATEAU (Théodore).

Traité complet des corps gras industriels. Paris, 1862. 12°.

Guide pratique de la connaissance et de l'exploitation des corps gras industriels, *etc.* Ed. 2. Paris, 1863. 8°.

CHAUMETON (François Pierre), J. CHAMBERET, & J. L. M. POIRET.

Flore médicale, peinte par Madame E. Panckoucke et par P. J. T. Turpin. Paris, 1814–20. 8 vols. 8°. (Also in fol.)

CHEETHAM (John).

On the Present Position and Future Prospects of the Supply of Cotton. *Journ. Soc. Arts*, xi. (1863) 255–268.

[A discussion opened by, on] "Cotton Cultivation and Supply." *Journ. Soc. Arts*, xvii. (1869) 651–653.

CHÉRUY-LINGUET (—.).

Ailanticulture pratique. Rapport présenté à la société d'agriculture, commerce, science et arts, du département de la Marne, le 15 Févr. 1869. Châlons-sur-Marne, 1869. 8°.

CHEVALLIER (Jean Baptiste Alphonse).

Notice sur le papier du riz . . . Extrait du *Journal des connaissances nécessaires et indispensibles*. Paris, 1840. 8°.

Sur les falsifications de la chicorée, dite café-chicorée. [*Annales d'hygiene publique*.] Paris, [1849]. 8°.

Dictionnaire des altérations et falsifications des substances alimentaires, médicamenteuses et commerciales, avec l'indication des moyens de les reconnaître. Paris, 1850–52. 2 vols. 8°. Ed. 2. 1853–55. Ed. 3. 1858. Ed. 4. 1875.

Notice historique sur l'opium indigène. [*Moniteur des hôpitaux*.] Paris, 1852. 8°.

Du café, son historique, son usage, son utilité, ses alterations, ses succédanés et ses falsifications, *etc.* Paris, 1862. 8°.

CHEVALLIER fils (Alphonse), & J. HARDY.

Manuel du commerçant en épicerie. Traité des marchandises qui sont du domaine de ce commerce ; falsifications qu'on leur fait subir ; moyen de les reconnaître. Paris, 1863. 12°.

Dictionnaire des substances alimentaires. Des falsifications qu'on leur fait subir, moyens de les reconnaître. Paris (Roret), 1874. 12°.

> The preceding with new title, and suppression of the co-author's name.

CHEVANDIER (Eugène).

Recherches sur l'emploi de divers amendements dans la culture des forêts. Paris, 1852. 4°.

CHIAPPONI (Agostino).

La silvicultura in Liguria, *etc.* Genova, 1876. 8°.

CHICCO (—.).

Cenni storici e statistici sulla coltivazione del sughero nell' Algeria. (Bollet. consol. public. per cura del Minist. degli affari esteri. xvi. (1880) fasc. 2.)

CHOMEL (Pierre Jean Baptiste).

Abrégé de l'histoire des plantes usuelles, *etc.* Paris, 1712. 8°. Ed. 2. 1715. 2 vols. Ed. 4. 1730. 3 vols. Ed. 5. 1738. Ed. 6. 1761. Ed. [7.] 1782. Ed. 7. [=8.] par J. B. N. Maillard, Beauvais et Paris, an xi. [1803]. 2 vols. For ed. 8 *see* DUBUISSON (J. R. J.).

CHRISTISON (*Sir* Robert).

Observations on a new variety of Gamboge from Mysore. *Pharm. Journ.* vi. (1846) 60–69.

On the Gamboge tree of Siam. [See also in *Proc. Royal Soc. Edinb.*] *Pharm. Journ.* x. (1850) 235–236.

On the properties of the ordeal-bean of Old Calabar, Western Africa. *Pharm. Journ.* xiv. (1855) 470–476.

On a new source of Kino. *Pharm. Journ.* xii. (1853) 377–379.

The effects of Cuca, or Coca, the leaves of Erythroxylon Coca. [*Brit. Med. Journ.*] *Pharm. Journ.* III. vi. (1876) 883–885.

CHRISTY (Thomas).

Forage plants, and their economic conservation by the new system of "Ensilage." London, 1877. 8°.

New Commercial Plants, with directions how to grow them to the best advantage. Nos. I.–IV. London, 1879–81. 8°.

CHURCH (A. H.). *See* JOHNSTON (J. F. W.).

CLARKE (Charles Baron).

Report on Cinchona·cultivation in Bengal. *Pharm. Journ.* III. i. (1870) 109–110.

Report on the Cultivation of Cinchona at Darjeeling for the half-year ending September 30, 1870. *Pharm. Journ.* III. i. (1871) 746.

CLARKE (Robert).

Notes on the Tallicoonah or Kundah oil. *Pharm. Journ.* ii. (1842) 341–342.

Some further remarks on the economical and medical uses of the oil commonly called Croupee on the Gold Coast, Touloucouna at the Gambia and. Senegal, and Kundah at Sierra Leone. *Pharm. Journ.* II. i. (1860) 540–542.

Observations on the Etua-tree (Kigelia africana). [Extract.] *Pharm. Journ.* II. iii. (1861) 182.

CLAVÉ (Jules).

Études sur l'économie forestière. Paris, 1862. 12°.

CLAYE (Louis).

Culture des fleurs et des plantes aromatiques. Fabrication des parfums en Portugal et dans ses colonies, *etc.* Paris, 1865. 8°.

CLAUSSEN (—. de).

On Papyrus, Bonapartea, and other plants which can furnish fibre for paper pulp. [Extract from *Brit. Assoc. Reports.*] *Pharm. Journ.* xv. (1855) 236–237.

On the Hancornia speciosa, artificial gutta-percha and india-rubber. [*Id.*] *Pharm Journ.* xv. (1855) 237–238.

On the employment of algae and other plants in the manufacture of soaps. [*Id.*] *Pharm. Journ.* xv. (1855) 238.

CLEGHORN (Hugh Francis Clarke).

Remarks on Calysaccion longifolium, *Wight. Pharm. Journ.* x. (1851) 597–598.

Memorandum upon the Paucontee, or Indian Gutta Tree of the Western Coast [Bassia elliptica, *Dalz.*]. Madras, 1858. 4°.

Report on the Forests of the Punjab and the Western Himalayas. Roorkee, 1864. 8°.

Observations on Forests and the Operations of the Forest Department. 1858. 8°.

Notes on the Agriculture of Malta and Sicily. Edinburgh, 1870. 8°.

The Forests and Gardens of South India. London, 1861. 12°.

CLOUET (J.), & J. DÉPIERRE.

Dictionnaire bibliographique de la garance [Rubia tinctoria]. Avec préface par J. Girardin. Paris, 1878. 8°.

COAZ (J.).

Die Kultur der Weide. Bern, 1879. 8°.

La culture des osiers. Traduit de l'allemand par A. Davall. Berne, 1879. 8°.

COBBETT (William).

The Woodlands, *etc.* London, 1825. 8°.

COCCONI (Giralamo).

Flora dei foraggi che spontanei o coltivati crescono nelle provincie Parmensi. Parma, 1856–60. 8°.

COHAUSEN (Johann Heinrich).

Neo Thea of nieuwe theetafel . . . ofte vel eene naaukeurige beschrijving van de krachten der in- en uitlandsche kruiden, bloemen, wortelen en planten om de zelve als thee te doen trekken . . . Vertaalt en met aantekeningen opgeheldert . . . door H. J. Grasper. Amsterdam, 1719. 8°.

Dissertatio . . . de Pica Nasi, *etc.* Amstelodami, 1716. 8°.

Satyrische Gedancken von der Pica Nasi, oder der Sehnsucht der luestern Nase, das ist; von dem heutigen Missbrauch- und schaedlichen Effect des Schnupf-Tabacks, *etc.* Leipzig, 1720. 8°.

COIN (Robert de).

History and Cultivation of Cotton and Tobacco. With map of North America. London, 1864. 8°.

COLLADON (Louis Théodore Frédéric).

Histoire naturelle et médicale des casses et particulièrement de la casse et de sénés employé en médicine. Montpellier, 1816. 4°.

> The botanical portion was the unavowed work of A. P. de Candolle.

COLLINGWOOD (Cuthbert).

On the British Lichens: their character and uses. *Pharm. Journ.* xviii. (1859) 362–365.

COLLINS (I. G.).

Scinde and the Punjaub . . . in respect to . . . the cotton markèts of the world, *etc.* Manchester, 1858. 8°.

COLLINS (James).

Notes on some new or little-known vegetable products. *Pharm. Journ.* II. xi. (1869) 66–70.

On India Rubber: its History, Commerce, and Supply. *Journ. Soc. Arts,* xviii. (1869) 80–92.

Materia medica notes. [*Journ. Bot.*] *Pharm. Journ.* III. ii. (1872) 1049.

On the study of Economic Botany, and its claims educationally and commercially considered. *Journ. Soc. Arts,* xx. (1872) 237–247. *Pharm. Journ.* III. ii. (1871) 691–695, 713–715, 737–739.

The India-rubber of commerce, with remarks on its preparation. [*Colonial News.*] *Pharm. Journ.* III. ii. (1871) 266–268.

Report on the Caoutchouc of commerce, information on the plants yielding it, their geographical distribution, and possible cultivation in India. London, 1872. 4°.

Museums: their commercial and scientific uses. [Singapore, 1874.] 8°.

COLLYER (Robert H.).

Rheea Fibre and Dr. Collyer's patent Rheea machine, with extracts on mode of cultivation by Dr. J. Forbes Watson. Calcutta, 1880. 8°.

COMENDATOR Y TELLEZ (Primo).

Estudio botánico, médico, farmacéutico y económico de las Solanáceas, seguido de una monografia de la Belladonna. Madrid, 1866. 4°.

COMES (Orazio).

Considerazioni sulla produzione del tabacco in Italia; e sulla convenienza di estendere la coltivazione. (*L'agricolt. meridionale*, Portici, (1881) Nos. 1–3.)

CONDAMY (A.).

Étude sur l'histoire naturelle de la truffe. Angoulême, 1876. 4°.

COOKE (Mordecai Cubitt).

The Seven Sisters of Sleep. Popular history of the seven prevailing narcotics of the world. London, [1860]. 8°.

On Pulu, and some analogous products of Ferns. *Pharm. Journ.* II. i. (1860) 501–504.

Kukui, or Kekune oil. [Aleurites triloba. From *Lond. Med. Rev.*] *Pharm. Journ.* II. ii. (1860) 282–283.

Pine wool. [From *The Technologist.*] *Pharm. Journ.* II. iii. (1861) 29–30.

Asafoetida in Affghanistan. A supplementary note. [From *The Technologist.*] *Pharm. Journ.* II. v. (1864) 583–584.

Hasan-i-Yusaf. [Isoëtes sp.] *Pharm. Journ.* III. i. (1870) 2–3.

Baobab. Adansonia digitata, L. *Pharm. Journ.* III. i. (1870) 64.

Spogel seeds. Plantago Ispaghula, *Roxb. Pharm. Journ.* III. i. (1870) 86.

Unto-mool. Tylophora asthmatica, *W. and A. Pharm. Journ.* III. i. (1870) 104–108.

Water Chestnuts. [Trapa.] *Pharm. Journ.* III. i. (1870) 125.

Gold Thread. Coptis tecta, *Wall.*, and Coptis trifolia, *Salisb. Pharm. Journ.* III. i. (1870) 161.

Medicinal Ferns. *Pharm. Journ.* III. i. (1870) 181, 204–205.

Guarana. Paullinia sorbilis, *Mart. Pharm. Journ.* III. i. (1870) 221.

Beech Morels. [Cyttaria.] *Pharm. Journ.* III. i. (1870) 264–265.

Kashmir Morels. [Morchella.] *Pharm. Journ.* III. i. (1870) 345–346.

Kali-kutki. (Picrorhiza Kurroa.) *Pharm. Journ.* III. i. (1870) 502–503.

Clearing nuts. (Strychnos potatorum.) *Pharm. Journ.* III. ii. (1871) 43–44.

Kafur kachri. (Hedychium spicatum.) *Pharm. Journ.* III. i. (1871) 603–604.

Jew's Ear. (Hirneola Auricula-Judae.) *Pharm. Journ.* III. i. (1871) 681.

Hermodactyls. *Pharm. Journ.* III. i. (1871) 784–785.

COOKE (Mordecai Cubitt) *continued* :—

Variability in the [medicinal] activity of leaves. *Pharm. Journ.* III. i. (1871) 861.

Ajwan or Omum. (Ptychotis Ajowan.) *Pharm. Journ.* III. i. (1871) 1007–1008.

Quinoa. (Chenopodium Quinoa.) *Pharm. Journ.* III. iii. (1872) 281.

Two medicinal barks from Ceylon. [Samadera bark, Samadera indica, *Gaertn.*, and Kokoon bark, Kokoona zeylanica, *Thwaites.*] *Pharm. Journ.* III. ii. (1872) 541.

Oriental aconite. *Pharm. Journ.* III. iii. (1873) 563–565.

Report . . . on the Gums, Resins, Oleo-resins, and Resinous Products in the India Museum, or produced in India. London, 1874. fol.

Report . . . on the Oil Seeds and Oils in the India Museum, or produced in India. London, 1876. fol.

Costus. *Pharm. Journ.* III. viii. (1877) 41–44.

> Bibliography of the subject on last page.

COOLEY (W. D.).

On the Cinnamon region of Eastern Africa. [From *The Athenæum.*] *Pharm. Journ.* ix. (1849) 34–36.

COOPER (E.).

Forest culture and Eucalyptus trees. San Francisco, 1876. 12°.

COOPER (H. Stonehewer).

Coral Lands. London, 1880. 2 vols. 8°.

> Contains much about the vegetable products of the South Pacific Islands.

COPE's Tobacco Plant ; *see* Periodicals.

CORDER (O.).

Note on Gentian root. *Pharm. Journ.* III. x. (1879) 221.

CORDIER (F. S.).

Guide de l'amateur de champignons, ou précis de l'histoire des champignons alimentaires, vénéneux et employés dans les arts, qui croisent sur le sol de la France. Paris, 1826. 12°. Ed. 2. 1836.

> In German. Quedlinburg, 1838. 8°.

CORENWINDER (Benjamin).

Influence de l'effeuillaison des betteraves sur le rendement et la production du sucre. Paris, [1876]. 8°.

Études comparatives sur les blés d'Amérique, de l'Océanie et des blés indigènes. Paris, [1876]. 8°.

CORENWINDER (Benjamin), & C. CONTAMINE.

De l'influence des feuilles sur la production du sucre dans les betteraves. Lille, 1879. 8°.

CORNAGGIA (Giovanni).

Nuovo sistema di coltivazione dei gelsi. Milano, 1879. 8°.

COSSIGNY DE PALMA (Joseph François Charpentier).

Essai sur la fabrique de l'indigo. Isle de France, 1779. 4°.

—————— Memoir . . . on the cultivation of Indigo. Calcutta, 1789. 4°.

COSSON (Ernest).

Note sur l'acclimatation de l'Eucalyptus Globulus. Paris, [1875]. 8°.

COSTE (Jean François), & Remi WILLEMET.

Essais botaniques, chimiques et pharmaceutiques sur quelques plantes indigènes substituées avec succès à des végétaux exotiques. Nancy, 1778. 8°. Ed. 2. (Matière médicale indigéne, etc.) 1793.

COTTA (Heinrich).

Abriss einer Anweisung zur Vermessung, Beschreibung, Schaetz-ung und forstwirthschaftl. Eintheilung der Waldungen. Dresden, 1815. 8°.

Systemat. Anleitung z. Taxation der Waldungen. Berlin, 1804. 8°.

Anweisung zur Forst-Einrichtung und Abschaetzung. Dresden, 1820. 8°.

Anweisung zur Waldbau. Dresden und Leipzig, 1817. 8°. Ed. 2. 1817. Ed. 3. 1821. Ed. 4. 1828. Ed. 5. von August Cotta, 1835. Ed. 6. 1845. Ed. 7. von Edm. von Berg, Leipzig, 1849. Ed. 8. 1856. Ed. 9. von Heinrich von Cotta, 1865.

Entwurf einer Anweisung zur Waldwerthberechnung. Dresden, 1818. 8°. Ed. 2. 1819. Ed. 3. 1840.

Grundriss der Forstwissenschaft. Dresden und Leipzig, 1831. 8°. Ed. 2. 1836–38; Beilage 1838. Ed. 3. "von seinen Soehnen," 1842–43. Ed. 4. 1849. Ed. 5. 1860. Ed. 6. von Heinr. und Ernst von Cotta, 1872.

Huelfstafeln fuer Forstwirthe und Forsttaxatoren. Dresden, 1821. 8°. Ed. 2. 1841.

Naturbeobachtungen ueber die Bewegung und Function des Saftes in der Gewaechsen, mit vorzuegl. Hinsicht auf Holzpflan-zungen. Weimar, 1806. 4°.

Tafeln zur Bestimmung des Inhalts der ruden Hoelzer, der Klafterhoelzer und der Reisigs, so wie zur Berechnung der Nutz- und Bauholzpreise. Dresden, 1816. 8°. Ed. 2. 1823; Nachtrag, 1824. Ed. 3. 1838. Ed. [4.?] 1841. · Ed. 4.

COTTA (Heinrich) *continued :*—

von August Cotta, 1845. Ed. 5. 1847. Ed. 6. 1851. Ed. 7. 1857. Ed. 9. 1859. Ed. 10. 1862. Ed. 11. von Heinrich von Cotta, 1864. Ed. 13. 1870. Ed. 14. 1874. Supplement zur 1. bis 14. Aufl., 1875. Ed. 15. 1878.

—— fuer de oesterreich. Staaten eingerichtet, von Leopold Grabner. Wien, 1827. 8°.

Die Verbindung des Feldbaues mit dem Waldbau, oder die Baum-wirthschaft. Dresden, 1819. 8°. 1e–3e Fortsetzung, 1820–22.

Krutsch und Reum, Ansichten der hoehern Forstwissenschaft, nach ihrem Wesen und Einfluss auf dem Staat. Herausgegeben von C. F. Schlenkert. Dresden, 1818. 4°.

Erlaeuterung der Forsteinrichtung durch ein ausgefuehrtes Beis-piel des Zugabe zu dessen Grundriss der Forstwissenschaft, und als 2r Thl. der Anweisung zur Forsteinrichtung und Abschaetz-ung. Dresden und Leipzig, 1832. 8°.

—— Huelfstafeln fuer Forstwirthe und Forsttaxatoren auf das oesterr. Mass reducirt. Zum Gebrauch fuer oesterr. Forst-maenner eingerichtet von S. Ph. Wander. Prag, 1833. 8°.

Anweisung zur Waldwerthberechnung. Ed. 4. von A. Cotta, Leipzig, 1849. 8°.

—— Fuer die Beduerfniss der k. k. oesterreichischen Staaten . . . von Heinrich von Cotta. Leipzig, 1854. Ed. 2. 1861. Ed. 3. 1866. Ed. 4. 1874.

COTTON (Joseph Gustave Stanilas).

Étude comparée sur le genre Krameria et les racines qu'il fournit à la medicine. Thèse. Paris, 1868. 4°.

COUBARD D'AULNAY (G. E.).

Monographie du café, ou manuel de l'amateur de café, ouvrage contenant la description et la culture du cafier, l'histoire du café, ses caractères commerciaux, sa préparation, et ses pro-priétés. Paris, 1832. 4°. Ed. 2. 1842.

COUTANCE (A.).

Histoire du chêne dans l'antiquité et dans la nature. Ses applications à l'industrie, aux constructions navales, aux sciences et aux arts, *etc.* Paris, 1873. 8°.

L'olivier, histoire, botanique, regions, culture. Paris, 1877. 8°.

COURBON (A.).

On the Taenifuges (Anthelmintics) of Abyssinia. [From the *Repert. de Pharm.*] *Pharm. Journ.* II. iii. (1861) 20–23.

Other medicinal plants of the Abyssinians. [Extract.] *Pharm. Journ.* II. iii. (1861) 23–24.

COZE (L.).

Histoire naturelle et pharmacologique des médicaments narcotiques fournis par le règne végétal. Thèse. Strasbourg, 1853. 4°.

CRACE-CALVERT (Frederick).

Abstract of a lecture on Caoutchouc and Gutta Percha. *Pharm. Journ.* xii. (1853) 423–426.

On Starches, the purposes to which they are applied, and improvements in their manufacture. *Journ. Soc. Arts,* viii. (1859) 87–93.

Dyes and dye-stuffs other than Aniline. Red colouring substances: — Madder. [Cantor Lecture, *Journ. Soc. Arts.*] *Pharm. Journ.* III. ii. (1871) 394–396, 414–417.

——— Munjeet; Campechy Peach, Sapan Cane and Bar woods; Alkanet Root; Safflower; Cochineal, Lac Dye; Murexide. *Pharm. Journ.* III. ii. (1871) 435–437, 454–457.

——— Blue colouring substances. — Indigo, Orchil, Cudbear, Litmus, Prussian Blue and Ultramarine. *Pharm. Journ.* III. ii. (1871) 513–515, 535–537.

——— [Yellow, etc.] Quercitron, Fustic, Persian Berries, Wild Aloes, Turmeric, Annatto, Ilixanthine, Lakao, Tannin matters, Gall-nuts, Sumach, Divi-divi, Myrobalans, Catechu. *Pharm. Journ.* III. ii. (1871) 537–538, (1872) 573–575.

CRAIG (William).

The medicinal plants of Scotland. [Extract.] *Pharm. Journ.* III. vii. (1877) 911–914.

CRAMER (Joh. Andreas).

Anleitung zum Forstwesen. Braunschweig, 1766. fol.

CRAUFURD (John).

On the migration of cultivated plants in reference to ethnology. [*Brit. Assoc. Report.*] *Journ. Bot.* iii. (1866) 317–332.

The Sugar Cane. [Extract.] *Pharm. Journ.* II. ix. (1867) 76–76.

CRAWFURD (John).

On the Cotton Supply. *Journ. Soc. Arts,* ix. (1861) 399–413.

CREASE (Orlando).

On Gum Anime. *Pharm. Journ.* vii. (1847) 15–16.

CREIGHTON (Benjamin T.).

The Culture of Tobacco in Ohio. [*Amer. Journ. Pharm.*] *Pharm. Journ.* III. vii. (1876) 27–28.

CROISETTE-DESNOYERS (L.).

Notice forestière sur les landes de Gascogne. Clermont, 1875. 8°.

CROOKES (William).

On the Manufacture of Beetroot Sugar in England and Ireland. London, 1870. 8°.

CROSS (Robert).

Report to the Under Secretary of State for India, on the Pitayo Chinchona, and on proceedings while employed in collecting Chinchona seeds in 1863. London, 1865. 8°.

Abstract of a report on the Pitayo Cinchonas. [*Gardeners' Chronicle.*] *Pharm. Journ.* II. vii. (1865) 120–121.

The India-rubber acclimatization experiment. [Extract.] *Pharm. Journ.* III. vii. (1876) 194–197.

Report on the investigation and collecting of plants and seeds of the Indiarubber trees of Para and Ceara, and Balsam of Copaiba. London, 1877. fol.

CRUEWELL (G. A.).

Liberian Coffee in Ceylon. The history of the introduction and progress of the cultivation. Colombo, 1878. 8°.

Coffee-cultivation, Brazil as a coffee-growing country; its capabilities, the mode of cultivation and prospects of extension. Colombo, 1878. 8°.

CULPEPER (Nicholas).

The English Physician, or an astrologo-physical discourse on the vulgar herbs of the nation. London, 1652. 12°.

The English Physician enlarged, *etc.* London, 1653. 8°. Other eds. in 1661, 1695, 1714, 1725, 1733, 1784, 1792. [1820?] "Crosby's improved edition," 1814. 12°. There are two Welsh versions; Caernarfon, 1818-19. 8°. and [1862?] 16°.

Pocket Companion to C.'s Herbal, *etc.* By J. Ingle. London, 1820. 12°.

CUNNINGHAM (D. D.).

Microscopical notes regarding the Fungi present in Opium blight. Calcutta, 1875. 8°.

CURRIE (John H.).

On two varieties of false Jalap. [From the *N. Y. Journ. Pharmacy.*] *Pharm. Journ.* xi. (1852) 521–523.

CURTIS (William).

A catalogue of the British medicinal, culinary, and agricultural plants, cultivated in the London Botanical Garden. London, 1783. 8°.

CUZENT (G.).

Tahiti. Considérations . . . botaniques sur l'île . . . végétaux susceptibles de donner des produits utiles au commerce et à

CUZENT (G.) *continued*: —

. l'industrie et de procurer des frets de retour aux navires. Cultures et productions horticoles. Catalogue de la flore de Tahiti. Rochefort, 1860. 8°.

Du Tacca pinnatifida, pia de Taïti, de sa fécule, de sa paille, et du Pandanus odoratissimus. Paris, 1861. 8°.

On Kava or Ava (Piper methysticum). [From *The Technologist.*] *Pharm. Journ.* II. iv. (1862) 85–87.

DAJI (Narayan).

A new Indian remedy. [Ailanthus excelsa, *Roxb.*] *Pharm. Journ.* III. i. (1870) 154–156, 175–176, 193–194.

DALE (Samuel).

Pharmacologia, seu manuductio ad materiam medicam. Londini, 1693. 12°. Supplementum, 1705. Ed. 2. 1737. 4°.

DALGAS (E.).

Anvisning til Anlaeg af Smaaplantninger omkring Gaarde og Haver samt til Anlaeg af levende Hegn og Anglaeg af Pile- culturer. Ed. 2. Kjoebenhavn, 1876. 8°.

Om Plantning i Jylland, navnlig i dets Hedeegne . . . af "Tids- skrift for Skovbrug," *etc.* Kjoebenhavn, 1878. 8°.

DALMON (J.).

Gigartina aciculáris as an adulterant of Carrageen Moss. *Pharm. Journ.* III. iv. (1874) 616.

DALZELL (Nicol Alexander).

On the Antiaris saccidora, or Sack-tree. [From *Hooker's Journ. Bot.*] *Pharm. Journ.* xi. (1851) 114–115.

DANGERS (—.).

Neue Gespinnstpflanzen (Abutilon Avicennae, Laportea pustulata, Apocynum cannabinum, Asclepias cornuta). (Fuehling's *Landw. Zeitung*, 1880, Heft 4).

DANIELL (William Freeman).

On the D'Amba, or Dakka, of Southern Africa. [Cannabis sativa, *var.*] *Pharm. Journ.* ix. (1850) 363–365.

On the Sanseviera guineensis, or African hemp. *Pharm. Journ.* xi. (1852) 130–132.

On the Zea Mays and other cerealia of Western Africa. *Pharm. Journ.* xi. (1852) 347–352, 395–401.

On the Synsepalum dulcificum, *De Cand.*; or, miraculous berry of Western Africa. *Pharm. Journ.* xi. (1852) 445.

On the Pterocarpus erinaceus, or Kino tree of West Africa. *Pharm. Journ.* xiv. (1854) 55–61.

DANIELL (William Freeman) *continued :—*

On the Habzelia aethiopica, Ethiopian or monkey pepper. *Pharm. Journ.* xiv. (1854) 112–116.

Kātemfe, or the miraculous fruit of Soudan. [Phrynium Danielli, *Bennett.*] *Pharm. Journ.* xiv. (1854) 158–160.

On the Cubeba Clusii of Miquel, the Black Pepper of Western Africa. *Pharm. Journ.* xiv. (1854) 198–203.

On the Amoma of Western Africa. *Pharm. Journ.* xiv. (1855) 312–318, 356–363; xvi. (1857) 465–472, 511–517.

On the Frankincense-tree of Western Africa. (Daniellia thurifera, *Bennett*). *Pharm. Journ.* xiv. (1855) 400–403, 463.

On Egusé oil, a new vegetable product from South Africa. *Pharm. Journ.* xvi. (1856) 307–309.

Some observations on the Copals of Western Africa. *Pharm. Journ.* xvi. (1857) 367–373, 423–426.

On Coelocline polycarpa, A. DC., the Berberine or yellow-dye tree of Soudan. *Pharm. Journ.* xvi. (1857) 398–401.

On a red Canella bark from the West Indies. *Pharm. Journ.* xviii. (1859) 503.

On African Turmeric. *Pharm. Journ.* II. i. (1859) 258–260.

On Ricinus inermis, *Will.* (var. manchuriensis). *Pharm. Journ.* II. iii. (1861) 15–16.

On the Cascarilla, and other species of Croton, of the Bahama and West India Islands. *Pharm. Journ.* II. iv. (1862) 144–150 [with two plates]. *Id.* 226–231 [with one plate].

On the production of Hydrocyanic acid from Bitter Cassava root. *Pharm. Journ.* II. vi. (1864) 302–304.

On the Kola-nut of tropical West Africa. (The Guru-nut of Soudan.) *Pharm. Journ.* II. vi. (1865) 450–457.

DARRU (Albert).

Manuel, avec calendrier agricole et horticole, du cultivateur algérien. Ouvrage fait d'après le programme de la société d'agriculture d'Alger. Saint-Loger, 1873. 8°.

DAVET (Gaston).

De quelques cholagogues nouveaux d'origine végétale. Paris, 1880. 8°.

DAYSON (F. A.).

Cassava Bread, prepared from the root of the Cassava plant. Janipha Manihot. (Euphorbiaceae.) *Pharm. Journ.* II. ii. (1860) 13–15.

DEBAY (Auguste).

Les influences du chocolat, du thé, et du cafè sur l'économie humaine, *etc.* Paris, 1864. 18°.

DEBEAUX (J. Odon).

Essai sur la pharmacie et la matière médicale des Chinois. Paris, 1865. 8°.

Notes sur quelques matières tinctoriales des Chinois. Paris, 1866. 8°.

DE BRUYN KOPS (G. F.).

Note on some of the productions of the Rhio-Lingga archipelago. *Pharm. Journ.* xvi. (1856) 43–44.

DE CAILUS (—.).

Histoire naturelle du cacao et du sucre. Ed. 2. Amsterdam, 1720. 8°.

DECAISNE (Joseph).

Recherches anatomiques et physiologiques sur la garance, sur le développement de la matière colorante dans cette plante, sur sa culture et sa préparation, suivies de l'examen botanique du genre Rubia et de ses espèces. Bruxelles, 1837. 4°.

DÉCOBERT (D.).

Culture du tabac. Lille, 1876. 8°.

DECHARME (C.).

De l'opium indigène extrait du pavot-oeillette, de l'identité de sa morphine avec celle de l'opium exotique, *etc.* Amiens, 1862. 8°.

DECKER (M.).

Die Kleeseide [Cuscuta]. Ein sehr gefaehrlicher Feind der Landwirthschaft, namentlich des Futterbaues. Luxemburg, 1880. 8°.

DECREPT (A.).

L'arbre vert en Picardie, son emploi avantageux dans le terrains calcaires et accidentés. Amiens, 1875. 8°.

DELABARRE (E. F. de), & E. CHAUME.

Améliorations chimiques et mécaniques apportées dans la fabrication et le raffinage du sucre de cannes et de betteraves. Paris, 1847. 8°.

DELCHER (E.).

Recherches historiques et chimiques sur le cacao et ses diverses préparations. Histoire, lieux où croît le cacao, culture, récolte et terrage, etc. . . . . Bibliographic des ouvrages sur le cacao, 1609 à 1830. Paris, 1873. 8°.

DELCHEVALERIE (G.).

Sur les végétaux d'ornement et d'utilité qui sont cultivés en Egypte. Paris, 1873. 8°.

Henna. [Extract.] *Pharm. Journ.* III. v. (1874) 8.

DELILE (Alire Raffeneau).

Dissertation sur les effets d'un poison de Java, appelé Upas tieuté, et sur la noix vomique, la fève de St. Ignace, le Strychnos potatorum, et la pomme de Vontac, qui sont du même genre de plantes, que l'Upas tieuté. Paris, 1809. 4°.

DELIOUX DE SAVIGNAC (—.).

The Myrtle and its properties. [*Repert. de Pharm.*] *Pharm. Journ.* III. vi. (1875) 346.

DELLA SUDDA (Georges).

Monographie des opiums de l'empire ottoman envoyés à l'Exposition universelle de Paris. Paris, 1867. 8°.

DELONDRE (August), & Apollinaire BOUCHARDAT.

Quinologie. Des quinquinas et des questions qui dans l'état présent de la science et du commerce s'y rattachent avec le plus d'actualité. Paris, 1854. 4°.

The Cinchona barks, and the more important questions which in the present state of science and commerce relate to them. [From the *Repert. de Pharmacie.*] *Pharm. Journ.* xiv. (1854) 77–83, 165–168, (1855) 513–517, 570–574.

DELTEIL (A.).

Étude sur la Vanille. Paris, 1874. 8°.

DEMAN (E. F.).

The Flax Industry; its importance and progress: also its cultivation and management, and instructions in the various Belgian methods of growing and preparing it for market, *etc.* London, 1852. 8°.

DEMERSAY (L. Alfred).

Du tabac du Paraguay; culture, consommation et commerce, avec une lettre sur l'introduction du tabac en France, par Ferdinand Denis. Paris, 1851. 8°.

Étude économique sur le Maté ou thé du Paraguay. Paris, 1867. 8°. *

DEMOOR (V. P. G.).

Traité du culture du tabac. Description historique, botanique et chimique. Luxembourg, [1853?]. 12°.

Lin, et des differents modes de rouissage. Bruxelles, 1855. 8°.

DEMONTZEY (P.).

Studien ueber die Arbeiten der Wiederbewaldung und Berasung der Gebirge . . . uebersetzt von Arth. von Seckendorff. Wien, 1880. 8°. Atlas. 4°.

DÉNIAU (C. C. Félix).

Le silphium (Assafoetida) précédé d'un mémoire sur la famille des Ombellifères considerée au point de vue économique, médical et pharmaceutique et d'observations sur les gommes résins. Paris, 1868. 8°.

DENNET (Charles F.).

Vegetable Fibres, with special reference to the textile fibres, Rhea or Ramie, Jute, New Zealand Flax,—their uses and abuse. Brighton, [1875]. 8°.

DESAGA (Oscar).

Ueber den Anbau des orientalischen Mohns, und Gewinnung des Opiums auf einheimischem Boden. Karlsruhe, 1868. 8°.

DESCHANELET-VALPÊTRE (J.).

Petite flore médicale illustrée, ou manuel des plantes les plus usitées dans le traitement des maladies, *etc.* Paris, 1862. 18°.

DESVAUX (Augustin Nicaise).

On Vanilla and its culture. [From the *Ann. Sci. Nat.*] *Pharm. Journ.* vii. (1847) 73–74.

DEVINCENZI (Giuseppe).

Della cultivazione del cotone in Italia. Londra, 1862. 8°.

On the cultivation of Cotton in Italy. London, 1862. 8°.

Della coltivazione del cotone in Italia. Torino, 1863. 8°.

DE VRIESE (Willem Hendrik).

De kampferboom van Sumatra, Dryobalanops Camphora, *Colebr.*, volgens Dr. F. Junghuhn's waarnemingen op de plaats zelve, en door nadere onderzoekingen toegelicht. Leiden, 1851. 4°.

On the Camphor-tree of Sumatra. [From *Hooker's Journ. Bot.*] *Pharm. Journ.* xii. (1852) 22–29.

Mémoire sur le camphrier de Sumatra et de Bornéo. Leide, 1857. 4°.

DE VRIJ (J. E.).

Cultivation of the Cinchona trees in Java. [Extract from a letter.] *Pharm. Journ.* II. ii. (1860) 220.

Note on Mr. Markham's Travels in Peru and India. [With remarks by D. Hanbury and others.] *Pharm. Journ.* II. iv. (1863) 439–441.

DE VRIJ (J. E.) *continued*:—

On the Cinchona bark of British India. *Pharm. Journ.* II. v. (1864) 593–599.

On the amount of alkaloids in the Cinchona trees cultivated in Java. *Pharm. Journ.* II. vi. (1864) 15–18.

On the possibility of manufacturing Neroli in the British colonies [Extract.] *Pharm. Journ.* II. vii. (1866) 477–478.

Samadera indica, *Gaertn. Pharm. Journ.* II. ii. (1872) 644–645, see also 654, 655.

Sale of Java bark at Amsterdam on 14th March, 1872. [With a note by J. E. Howard.] *Pharm. Journ.* III. ii. (1872) 945.

Presumed Hybrid between Cinchona Calisaya and C. succirubra. [With a note by J. E. Howard.] *Pharm. Journ.* III. v. (1874) 42.

A fast growing Cinchona which produces much Quinine. ["Cinchona pubescens, var. —, Howard."] *Pharm. Journ.* III. viii. (1878) 805.

DEXTER (Thomas E.).

Animal and Vegetable substances used in the Arts and Manufactures, illustrative of the Imports and Exports of Great Britain and her colonies, and explanatory of Dexter's Cabinet of objects. London, 1857. 8°.

DEY (Kanny Loll).

The indigenous Drugs of India; or short descriptive notices of the medicines, both vegetable and mineral, in common use among the natives of India. Calcutta, 1867. 8°.

DICKSON (James Hill).

A series of letters on the improved mode in the cultivation and management of flax, *etc.* London, 1846. 8°.

The Fibre Plants of India, Africa, and our Colonies. A Treatise on Rheea, Plantain, Pine Apple, Jute, African and China Grass, and New Zealand Flax (Phormium tenax), *etc.* London, [1863]. 8°.

DIEHL (C. Lewis).

American drugs. [*Proc. Amer. Pharm. Assoc.*] *Pharm. Journ.* III. i. (1871) 705–706.

DIERBACH (Johann Heinrich).

Abhandlung ueber die Arzneikraefte der Pflanzen verglichen mit ihrer Structer und ihren chemischen Bestandtheilen. Lemgo, 1831. 8°.

Grundriss der allgemeinen oekonomisch-technischen Gewaechse, oder systematische Beschreibung der nutzbarsten Gewaechse aller Himmelsstriche. Heidelberg, 1836–39. 2 vols. 8°.

DIERBACH (Johann Heinrich) *continued* :—
  Synopsis materiae medicae, oder Versuch einer systematischen
    Aufzaehlung der gebraeuchlichsten Arzneimittel. Heidelberg,
    1841–42. 8°.
DIETRICH (Albert).
  Handbuch der pharmaceutischen Botanik. Ein Leitfaden zu
    Vorlesungen und zum Selbsstudium. Berlin, 1837. 8°.
DIETRICH (David Nathaniel Friedrich).
  Forstflora oder Abbildungen und Beschreibung der fuer den Forst-
    mann wichtigen wildwachsenden Baeume und Straeucher,
    welche in Deutschland wildwachsen sowie der auslaend-
    ischen daselbst im Freien ausdauernden. Jena, 1828–33. 8°.
    Ed. 2. 1838–40. 4°. Ed. 4. Leipzig, 18[63–]67. Ed. 5.
    Dresden, 18[74–]80.
  Flora medica oder Abbildung der wichtigsten officinellen Pflanzen,
    etc. Jena, 1831. 4°.
  Das wichtigste aus dem Pflanzenreiche, fuer Landwirthe, Fabrik-
    anten, Forst- und Schulmaenner. Jena, 1831–38. 4°. Ed. 2.
    1840.
  Taschenbuch der Arzneigewaechse Deutschlands. Jena, 1838.
    8°.
  Deutschlands oekonomische Flora oder Beschreibung und Abbild-
    ung aller fuer Land- und Hauswirthe wichtigen Pflanzen.
    Jena, 1841–43. 3 vols. 8°. (Vol. i. Die Futterkraeuter.
    ii. Die Unkraeuter. iii. Getraidearten, Oelgewaechse, *etc.*)
  Taschenbuch der pharmaceutisch-vegetabilischen Rohwaerenkunde
    fuer Aerzte, Apotheker und Droguisten. Jena, 1842–46.
    2 vols. 8°.
DIETRICH (Friedrich Gottlieb).
  Der Apothekergarten, oder Anweisung in Apotheken brauchbare
    Gewaechse zu erziehen. Berlin, 1802. 8°.
DIEU (Alphonse).
  Histoire du Curare. Thèse. Strasbourg, 1864. 4°.
DIOSZEGI (Samuel).
  Orvosi füvészkönyv mint a magyar füvészkönyv praktika része.
    (Medicinische Botanik). Debrecin, 1813. 8°.
DIOT-GILMAT (—.).
  Culture du houblon. [Humulus Lupulus.] Le Mans, 1875.
    8°.
DITRICH (Ludwig).
  Plantae officinale indigenae, linguis in Hungaria vernaculis
    deductae. Budae, 1835. 8°.

DITTWEILER (Wilhelm).

Lehrbuch der Botanik fuer Thieraerzte, Landwirthe und Pharmaceuten und die betreffenden Lehranstalten zum Gebrauch
bei Vorlesungen und zum Selbstunterricht. Stuttgart, 1847.
8°.

DIVE (Hippolyte).

Monographie industrielle et commerciale du pin maritime. Mont-
de-Marsan, 1864. 8°.

DIXON (F.).

The Melilotus caerulea. [See also PEREIRA.] Pharm. Journ. ii.
(1843) 463, 464.

DOASSANS (E.).

Étude botanique, chimique et physiologique sur le Thalictrum
macrocarpum. Paris, 1881. 8°.

DOBEL (Karl Friedrich).

Synonymisches Woerterbuch der in der Arzneikunde und im
Handel vorkommenden Gewaechse, etc. Kempten, 1830. 8°.

DOEBNER (E. Ph.).

Lehrbuch der Botanik fuer Forstmaenner, nebst einem Anhang;
Die Holzgewaechse Deutschlands und der Schweiz, etc. As-
chaffenburg, 1853. 8°. Ed. 2. 1858. Ed. 3. 1865.

DOUGLAS (J.).

Notes on Indian roses and their products. [Gardeners' Chronicle.]
Pharm. Journ. III. viii. (1878) 811.

DOUGLAS (James).

Arbor yemensis, fructum Cofé ferens: or a description of the
coffee tree. London, 1727. fol. Supplement, 1727.

DOUMET-ADANSON (N.).

Sur les fôrets de la Corse, et la destruction déplorable des laricios
archiséculaires qu'elles renferment. Paris, [1874]. 8°.

DORAT (Charles).

On the production of Balsam of Peru. [Extract from a letter.]
Pharm. Journ. II. ii. (1860) 172–174.

DOWDESWELL (G. F.).

Another report of the properties of the Coca leaf. [Extract.]
Pharm. Journ. III. vi. (1876) 946–947.

[DRAEBYE (Frants).]

Forsoeg til Hoer-Avlingen og dens Tilberedelse efter den Hol-
landske, Irlandske og Scotske Maade. Kjoebenhavn, 1778. 8°.

DRAPER (Harry Napier).

Pharmacy at the Dublin Exhibition. [Colonial products, etc.]
Pharm. Journ. II. vii. (1865) 147–152.

DREJER (Salomon Thomas Nicolai).
Compendium i den mediciniske Botanik. Kioebenhavn, 1840. 8°.
Anvisning til at Kjende de danske Foderurter. Ed. 3. Kioeben-
havn, 1847. 8°.

DRESSLER (E.).
Die Weisstanne (Abies pectinata) auf dem Vogesensandstein.
Ein Wort zur Anregung fuer deren moeglichst ausgedehnte
Verbreitung auf aehnlichen Standorten. Strassburg, 1880. 8°.

DROUET (L. J.).
L'arboriculture demontrée en pratique, *etc.* Vernon, 1873. 8°.

DRUMMOND (A. T.).
Canadian Timber Trees, their distribution and preservation.
Montreal, 1879. 8°.

DRURY (Heber).
The useful plants of India alphabetically arranged, with botanical
descriptions, vernacular synonyms, and notices of their economical value in commerce, medicine, and the arts. Madras,
1858. 8°. Ed. 2. London, 1873.

DUBOIS (Fr.).
Matière médicale indigène, ou histoire des plantes médicales qui
croissent spontanément en France et en Belgique. Tournai,
1848. 8°.

DU BREUIL (Alphonse).
Cours élémentaire théorique et pratique d'arboriculture, *etc.*
Paris, 1846. 8°. Ed. 2. 1850. Ed. 4. 1857-58. Ed. 6.
1867-72. 2 vols. Ed. 7. 1875-76. 12°.
Manuel d'arboriculture des ingénieurs. Paris, 1860. 12°.
The scientific and profitable culture of Fruit-trees . . . adapted for
English cultivators by W. Wardle, *etc.* Ed. 2. London, 1872. 8°.
Cours d'arboriculture. Les vignobles et les arbres à fruits à cidre.
L'olivier. Le noyer. Le mûrier et autres espèces économiques.
Paris, 1874. 12°.

DUBREUIL (H.).
Histoire naturelle et médicale de quelques végétaux de la famille
des Euphorbiacées. Thèse. Paris, 1835. 4°.

DUBRUNFAUT (—.).
Art de fabriquer le sucre de betteraves. Paris, 1825. 8°. *

DUBRUNFAUT (Augustin Pierre).
Sucrage des vendanges avec les sucres raffinés de canne, de
betterave, *etc.* Paris, 1854. 8°. Ed. 2. 1854. [Ed. 3?] 1874.
Le sucre dans ses rapports avec la science, l'agriculture, l'industrie,
le commerce, l'économie publique et administrative, *etc.* Paris,
1873. 8°.

DUBUISSON (J. R. Jacquelin).

Plantes usuelles indigènes et exotiques. Paris, 1809. 2 vols.
8°. *See also* P. J. B. CHOMEL *for previous issues.*

DUCHASSAING (Pierre).

The bark of the Adansonia digitata, or Baobab tree, a substitute
for Cinchona bark. [From the *Journ. de Pharmacie.*] *Pharm.
Journ.* viii. (1848) 89.

DUCHESNE (Antoine Nicolas).

Manuel de botanique, contenant les propriétés des plantes utiles
pour la nourriture, d'usage en médicine, employées dans les
arts, d'ornement pour les jardins, *etc.* Paris, 1764. 8°.
[*Anon.*]

DUCHESNE (Édouard Adolph).

Traité du maïs, ou blé de Turquie, contenant son histoire, sa
culture et ses emplois en èconomic domestique et en médicine.
Paris, 1833. 8°.

Répertoire et atlas des plantes utiles et des plantes vénéneuses du
globe, contenant . . . l'indication de leurs usages en médicine
humaine, en mèdicine vétérinaire, en économie domestique
et rurale, et dans les arts ou l'industrie. Paris, 1836. 2 vols.
8°. Bruxelles, 1846.

DUCHESNE (J. B.).

Guide de la culture des bois ou herbier forestier. Paris, [1825].
8°. Atlas fol.

DUCKWORTH (Dyce).

The Narthex Asafoetida; a description of the plant, its properties, and uses. *Pharm. Journ.* xviii. (1859) 464–468.

DUDGEON (J.).

The Ailanthus glandulosa in dysentery. [*Med. Times and Gazette.*]
Pharm. Journ. III. vii. (1876) 372–373.

DUFOUR (Louis).

Cours élémentaire sur les propriétés des végétaux et leurs applications. Neuchâtel, 1855. 8°.

Propriétés des végétaux et leurs applications à l'alimentation, la
mèdicine, la teinture, l'industrie. Neuchâtel, 1861. 8°.

DUFOUR (Philippe Sylvestre).

Traitéz nouveaux et curieux du café du thé du chocolate. Lyon,
1685. 12°. Ed. 2. 1688. Ed. 3. La Haye, 1693.
Other versions are extant in Latin, English, and Spanish.

DUFRESNE (Pierre).

Histoire naturelle et médicale de la famille des Valérianées.
Montpellier, 1811. 4°.

DUHAMEL DU MONCEAU (Henri Louis).

Traité des arbres et arbustes, qui se cultivent en France en
pleine terre. Paris, 1755 (also dated 1785). 2 vols. 4°.
Ed. 2. Paris, 1801–19. 7 vols. fol.

DUMAS (A.).

La culture maraichère. Traité pratique pour le midi, le centre
de la France et pour l'Algérie. Paris, 1880. 8°.

DUNAL (Michel Félix).

Histoire naturelle, médicale, et économique des Solanum, *etc.*
Montpellier, 1813. 4°.

DUNCAN (Andrew).

Catalogue of medical plants according to their natural orders.
Edinburgh, 1826. 8°.

DUPLESSY (F. S.).

Des végétaux résineux tant d'indigènes qu'exotiques, ou description complète des arbres, arbrisseaux, arbustes et plantes qui
produisent des résins, *etc.* Paris, 1802. 4 vols. 8°.

DUPONT (Adolphe E.).

Les essences forestières du Japon. Nancy, 1880. 8°.

Notes relatives aux Ka-kis cultivés japons. Toulon, 1880. 8°.

DUPONT (Adolphe E.), & Amédée BOUQUET DE LA GRYE.

Les bois indigénes et étrangers. Physiologie, culture, production,
qualités, industrie, commerce. Paris, 1875. 8°.

DUPUIS (—.).

The rose harvest in Adrianople. [Extract from Consular Report.]
Pharm. Journ. III. iv. (1873) 426.

DUPUIS (Aristide).

Traité élémentaire des champignons comestibles et vénéneux.
Paris, 1854. 12°.

Plantes agricoles et forestières. Ouvrage donnant la description,
la culture, et les usages des végétaux dont il traite. Paris,
1867. 8°.

Le règne végétal, devisé en traité de botanique générale, flore
médicale et usuelle, horticulture botanique . . . . par A. D.,
Fr. Gerard, O. Reveil et F. Herincq, *etc.* Paris, [1864–69].
9 vols. 8°., and 8 vols. 4°.

D'URBAN (W. S. M.).

English Ink-galls. [Abstract.] *Pharm. Journ.* II. iv. (1863)
520.

DURDEN (—.).

New materials for the manufacture of paper. [Extract from the
Leeds Intelligencer.] *Pharm. Journ.* xv. (1855) 30–32.

DUREAU (B.).

Sur la culture de la canne et fabrication du sucre en Louisiane. Paris, 1852. 8°.

La question des sucres devant le consommateur. Paris, 1863. 8°. Lond. 1864. 8°.

De la fabrication du sucre de betterave, dans ses rapports avec l'agriculture et l'alimentation publique, avec des considerations sur la partie économique et la legislation de cette industrie. Paris, 1858. 8°.

DU RONDEAU (—.).

Quelles sont les plantes les plus utiles des Pays-bas, et quel est leur usage dans la mèdicine et dans les arts? Bruxelles, 1772. 4°.

DUROSELLE (E.).

Le mélilot de Siberie dans les sols stériles. Nancy, 1873. 8°.

DUŠEK (Ignaz).

Anleitung zur Kultur des Faerberknoeterichs [Polygonum tinctorium] in Mittel-Europa und zur hierlaend. Indigobereitung. Frag, 1872. 16°.

DUTRONE DE LA COUTURE (Jacques François).

Traité de l'olivier. Aix, 1786. 2 vols. 8°.

Histoire de la canne et précis sur les moyens d'en extraire le sel essentiel, suivie de plusieurs mémoires sur le sucre, sur le vin de canne, sur l'indigo, sur les habitations, et sur l'état actuel de Saint-Domingue. Paris, 1790. 8°. Ed. 2. 1791. Ed. 3. 1801.

DUVE (J. D.).

Anweisung zum Anbau der behacten Brachfruechte oder Futtergewaechse. Celle, 1830. 8°.

DWYER (Hugo).

The preservation of herbs, roots, barks, etc., for pharmaceutical uses. [Extract.] *Pharm. Journ.* III. iv. (1874) 891.

DYER (William Thiselton).

On the medicinal properties of the Solanum tuberosum. *Pharm. Journ.* i. (1842) 590–591.

DYER (William Turner Thiselton).

On the plant yielding Latakia Tobacco. *Linn. Soc. Journ., Botany,* xv. (1876) 246–247.

> In the Corrigenda, p. ix, the spelling of the name is altered to Lattakia.

DYMOCK (W.).

Asafoetidas of the Bombay market. *Pharm. Journ.* III. v. (1875) 945.

DYMOCK (W.) *continued :*—

Note on Myrrh and its allied Gum resins. *Pharm. Journ.* III. vi. (1876) 661.

Notes on Indian drugs. *Pharm. Journ.* III. vi. (1876) 1002–1003, vii. 3–4, 109–110, 170–172, 190, 309–310, 350–351, 450–452, 491–493, 549–550, (1877) 729–731, 977–979, viii. 23–25, 101– 105, 161–162, 383–386, 483–485, 521–522, 564–566, 745– 747, 1001–1003, ix. (1878) 145–146, (1879) 894–896, 1015– 1017, 1033–1035, x. 121–123, 281–282, 381–383, 401–403, 461– 463, (1880) 581–582, 661–662, 829–831, 993–994, xi. 21–22, 169–170.

The botanical source of Sarcocolla. *Pharm. Journ.* III. ix. (1879) 735–736.

Indian Henbane. *Pharm. Journ.* III. xi. (1880) 369.

EATWELL (W. C. B.).

Selections from the Records of the Bengal Government, on the System of cultivating the Poppy and of preparing Opium in the Benares Opium Agency, *etc. Pharm. Journ.* xi. (1851) 269–271, (1852) 306–311, 359–364.

EBBINGHAUS (Julius).

Die Pilze und Schwaemme Deutschlands. Mit besondrer Rueck- sicht auf die Andwendbarkeit als Nahrungs- und Heilmittel, sowie auf die Nachtheile derselben. Leipzig, 18[62–]68. 4°.

EBERMAIER (Johann Christoph).

Vergleichende Beschreibung derjenigen Pflanzen, welche in den Apotheker leicht mit einander verwechselt werden. Braun- schweig, 1794. 8°.

Ueber die nothwendige Verbindung der systematischen Pflanzen- kunde mit der Pharmacie, und ueber die Bekanntmachung der giftartig wirkenden Pflanzen. Hannover, 1796. 8°.

Von den Standorten der Pflanzen im Allgemeinen und denen der Arzneigewaechse besonders. Muenster, 1802. 8°.

EBERMAIER (Carl Heinrich).

Plantarum Papilionacearum monographia medica. Berolini, 1824. 8°.

EEDEN (F. W. van).

Hortus batavus. Korte beschrijving van in- en uitheemsche planten, heesters en boomen, die voor de Nederlandsche tuinen kunnen worden aanbevolen. Amsterdam, 1868, 4°.

Issued in parts as, Bloemkundig woordenboek, *etc.*

EEDEN (F. W. van) *continued :—*

Algemeene beschrijvende catalogus der houtsoorten van Nederlandsche Oost-Indië, aanwezig in het kolonial museum, op het
paviljoen te Haarlem. Haarlem, [1872]. 8°.

Kolonial museum op het paviljoen bij Haarlem. Overzigt van
het museum als leiddrad voor de bezoekers. Haarlem, 1875.
8°.

EHRHART (Balthasar).

Unterricht von einer zu verfassenden Historie der nuetzlichsten
Kraeuter, Pflanzen und Baeume. Memmingen, 1752. 4°.

Oeconomische Pflanzenhistorie nebst dem Kern der Landwirthschaft, Garten- und Arzneikunst. Ulm und Memmingen,
1753–62. 12 vols. 8°.

EICHELBERG (Johan Friedrich Andreas).

Naturgetreue Abbildungen und ausfuehrliche Beschreibungen
aller in- und auslaendischen Gewaechse, welche die wichtigsten Produkte fuer Handel und Industrie liefern, als natur-
geschichtliche Begruendung der merkantilischen Waarenkunde.
Zuerich, 1845. 8°.

EICHLER (August Wilhelm).

Syllabus der Vorlesungen ueber Phanerogamenkunde. Kiel,
1876. 8°. Ed. 2. (Syllabus der Vorlesungen ueber specielle
und medicinisch-pharmaceutische Botanik.) Berlin, · 1880.
8°.

EIMBCKE (George).

Flora Hamburgensis pharmaceutica, oder Verzeichniss und Beschreibung der um Hamburg und in den angraenzenden
Laendern wildwachsenden Arzneipflanzen. Hamburg, 1822.
8°.

ELIA (Ferdinando d').

Istruzioni per ben coltivari i gelsi secondo le pratiche piu ricevuto,
con la giunta dei metodi in uso presso i Chinesi. Napoli,
1869. 8°.

ELLIOTT (Robert H.).

The Experiences of a Planter in the Jungles of Mysore. London,
1871. 2 vols. 8°.

Cardamom cultivation in Mysore. [Extract from the preceding.]
Pharm. Journ. III. viii. (1878) 547–548.

ELLIS (John).

An historical account of Coffee ; with an engraving and botanical
description of the tree ; to which are added sundry papers
relative to its use. London, 1774. 4°.

ELLIS (John) *continued :*—

A Description of the Mangosteen and the Bread Fruit, *etc.* London, 1775. 4°.

ELLISON (Thomas).

Handbook of the Cotton trade; or, a glance at the past history, present condition, and future prospects of the cotton commerce of the world. London, 1858. 8°.

ELLRODT (Theodor Christian).

Schwamm-Pomona, oder Beschreibung der essbaren und giftige Schwaemme Deutschlands. Baireuth, 1800. 12°.

ELMIGER (Joseph).

Histoire naturelle et médicale des Digitales. Montpellier, 1812. 4°.

ÉLOFFE (Arthur).

L'ortie, ses propriétés alimentaires, médicales, agricoles, et industrielles. Paris, [1862]. 18°.

Les champignons comestibles et vénéneux; guide pour les reconnaître. Paris, 1880. 16°.

ELST (S. L. W. van der).

Staats-koffiecultuur op Java. 's Gravenhage, 1874. 8°.

EMERSON (George B.).

A Report on the trees and shrubs growing naturally in the . forests of Massachusetts. Boston, 1846. 8°. Ed. 2. 1875. 2 vols.

The first edition was anonymous.

ENGEL (Louis Charles).

Histoire naturelle et pharmacologie des médicaments astringents végétaux. Thèse. Strasbourg, 1853. 4°.

Influence des climats et de la culture, sur les propriétés médicales des plantes. Thèse. Strasbourg, 1860. 4°.

ERNST (Adolph).

On the plants cultivated or naturalised in the valley of Caràcas, and their vernacular names. *Journ. Bot.* iv. (1867) 264–275, 287–296.

On Sabadilla. [Abstract of paper read before the Linnean Society.] *Pharm. Journ.* III. i. (1870) 513–514.

Die Betheiligung der vereinigten Staaten von Venezuela an der wiener Welt-Ausstellung, 1873. Caràcas, 1873. 8°.

Die Produkte Venezuelas auf der internationalen landw. Ausstellung in Bremen, 1874.

Coffee disease in New Grenada. *Nature*, xxii. (1880) 292.

ESCARPIT (J. A.).

Des plantations et des grandes arbres dans la Gironde et des départements limitrophes. Bordeaux, 1879. 16°.

ESPT (V. van der).

Hydrastis canadensis, or Golden Seal, and its alkaloids. [Extract.] *Pharm. Journ.* III. iii. (1873) 604.

EVANS (H. Sugden).

On the microscopic structure of Pepper and its adulterants. *Pharm. Journ.* II. i. (1860) 605–607, ii. 7–11.

On the Scammony and Jalap of commerce. *Pharm. Journ.* II. x. (1868) 220–223.

EVANS (W. J.).

The sugar planter's manual; being a treatise on the art of obtaining sugar from the sugar cane. London, 1847. 8°. Philadelphia, 1848.

EVELYN (John).

Silva; or, a discourse of forest trees and the propagation of timber in his Majesty's dominions. London, 1664. fol. Ed. 2. 1670. Ed. 3, 1679. Ed. [4?], 1706. Ed. 5. 1729. Other eds., York, 1776. 4°. 1786. 2 vols. 1801, and 1812. London, 1825.

EVERS (B.).

Indian medicinal plants. [Extract.] *Pharm. Journ.* III. v. (1875) 1028–1030.

EWART (J.).

Notes on Cinchona cultivation in British Sikkim. Calcutta, 1868. 8°.

FABER (August).

On Curaçoa Aloes. [From the *Central Blatt.*] *Pharm. Journ.* vii. (1848) 547.

FAIRHOLT (Frederick William).

Tobacco: its history and associations; including an account of the plant and its manufacture, *etc.* London, 1859. 8°. Ed. 2. 1876.

FALCONE (Pietro).

Istruzione pratica sulla coltivazione della zucchero, ossia canna da zucchero della China nell' India. Milano, 1865. 12°.

FALCONER (Hugh).

Some account of Aucklandia, a new genus of Compositae, believed to produce the Costus of Dioscorides. *Trans. Linn. Soc.* xix. [1842?] 23–31.

FALCONER (Hugh) *continued :—*
On Aucklandia Costus. *Pharm. Journ.* iii. (1844) 401–402.
Report on the Teak forests of the Tenasserim provinces, with other papers on the Teak forests of India and on the Tea Plant of Sylhet. Calcutta, 1852. 8°.
Report on Forest Administration in Burmah. Calcutta, 1864. 8°.

FAIRGRIEVE (Thomas).
Notes on the cultivation and preparation of Lactucarium. *Pharm. Journ.* III. iii. (1873) 972–973.

FARRE (Fred. J.).
On the growth and preparation of Rhubarb in China. *Pharm. Journ.* II. vii. (1866) 375–379.

FAUJAS-DE-SAINT-FOND (Barthèlemy).
Mémoire sur le Phormium tenax, improvement appelé lin de la Nouvelle-Zélande. Paris, 1813. 4°.

FAVIER (A.).
Les orties textiles (Ramie de Chine, etc.); histoire, culture, decortication. Paris, 1880. 12°.

FAVIER (P. A.).
Nouvelle industrie de la ramie. Notice sur la découverte de procédés mecaniques et chimiques pour la préparation et l'utilisation des fibres de la ramie, plante textile produisant une fibre plus forte que le lin et le chanvre, plus fine que le coton et la laine, et aussi brillante que la soie, *etc.* Paris, 1881. 8°.

FEKETE (Lajos).
A vörösfengő törzsek görbeségének oka. [The causes of the torsion in Larch trunks.] *Erdészeti lapok,* Budapest, (1880) 337–348.

FENNBRESQUE (S. P.).
Mémoire sur la destruction de la cuscute. Montauban, 1866. 8°.

FENZL (Eduard).
Bericht ueber einige der wichtigsten botanischen Ergebnisse der Bereisung der portugiesischen Colonie von Angola in Westafrica in den Jahren 1850–1860 durch Herrn Friedrich Welwitsch. Wien, 1863. 8°.

FÉRAND (Béranger).
The collection of Gum Senegal in Senegambia. [Extract.] *Pharm. Journ.* III. iv. (1873) 166–167.

FERGUSON (William).

    The Palmyra Palm of Ceylon. 1850. 8°.

FERMOND (Charles).

    Monographie du tabac, comprenant l'historique, . . . la description des principales espèces, *etc.* Paris, 1857. 8°.

FERO (Adolph).

    On the kinds of Rhubarb at present in Russian Commerce. [Extract.] *Pharm. Journ.* II. ix. (1867) 246-249.

FERRAGE (H.).

    Culture du lin et prèparation de la plante textile. Mémoire destiné à l'agronome qui voudra se livrer d'une manière progressive à la culture du lin. Toulouse, 1877. 8°.

FERRARI (Giovanni Battista).

    Hesperides, sive de malorum aureorum cultura et usu libri quatuor. Romae, 1646. fol.

FÈVRE (Justin Louis Pierre).

    Le tabac. Paris, 1863. 12°.

FIELD (H.).

    Agricultural Chemistry applied to the culture of the sugar cane, and relative estimation of the soils of British Guiana. Demerara, 1848. 8°.

FIELDER (C. H.).

    On Tea Cultivation in India. *Journ. Soc. Arts,* xvii. (1869) 291-300, 323-329.

FIGUEIREDO (Jeronymo Joaquim de).

    Flora pharmaceutica e alimentar portugueza, ou tractato daquelles vegetaes indigenas de Portugal, e outros nelle cultivados, cujos productos são usados, ou susceptiveis de se usar come remedios e alimentos. Lisboa, 1825. 8°.

FILET (G. J.).

    Plantkundig woordenboek van Nederlansche-Indië, met korte aanwijzingen van het geneeskundig- en huishoudelijk gebruik der planten, en vermelding der verschillende inlandsche en wetenschappelige benamingen. Leiden, 1876. 8°.

FILHOL (Édouard).

    On Indian Copal. [From the *Pharm. Centr. Blatt.*] *Pharm. Journ.* ii. (1843) 773-774.

    Note on some vegetable colouring matters. [From *Journ. de Pharm.*] *Pharm. Journ.* II. ii. (1860) 333-335.

FINCK (Hugo).

    Cultivation of Cinchona in Mexico. [Extract from a letter.] *Pharm. Journ.* III. i. (1870) 146.

FISCALI (Ferdinand).

Deutschlands Forstculturpflanzen. Ed. 2. Ollmuetz, 1858. 8°.

FISCHBACH (H.).

Katechismus der Forstbotanik. 2. gaenzlich umgearbeitete Auflage
. . . von J. V. Massaloup. Leipzig, 1862. 8°.
    Forms No. 6 of Weber's Illustrirte Katechismen.

FISCHER (Rudolf).

Die Feldholzzucht. Ein Beitrag zur Frage : Auf welche Weise
Kann sich der Besitzer eines grossen oder kleinen Gutes das
benoethigte Holz selbst produciren ? Mit besonderer Beruecksichtung der Korbweiden und Eichenschaelwaldanlagen. Berlin,
1878. 8°.

FLEISCHMANN (C. L.).

Memorial in relation to the manufacture of Beet Sugar. Washington, 1839. 8°.

FLEMING (John).

A Catalogue of Indian Medicinal Plants and drugs, with their
names in the Hindustani and Sanscrit languages. Calcutta,
1810. 8°.

FLICHE (H.).

Manuel de botanique forestière. Botanique anatomique et
physiologique, géographie botanique, botanique descriptive,
principales espèces forestières de France. Nancy, 1874. 8°.

FLICHE (P.), & L. GRANDEAU.

Recherches chimiques sur la végétation forestière. [Extract.]
Paris, 1878. 8°.

FLINT (Charles L.).

A practical treatise on grasses and forage plants, etc. New York,
1857. 8°. Ed. [3 ?] Boston, 1874.

FLUECKIGER (Friedrich A.).

Lehrbuch der Pharmacognosie des Pflanzenreiches. Naturgeschichte der wichtigeren Arzneistoffe vegetabilischen Ur-
sprungs. Berlin, 1867. 8°. Ed. 2. 1880.

Note on a new kind of Kamala. [With a note by D. Hanbury.]
Pharm. Journ. II. ix. (1867) 279-282. *See also* pp. 310-312
(1868).

Gummi und Bdellium vom Senegal. (Aus der *Schweizerischen
Wochenschrift fuer Pharmacie.*) Schaffhausen, 1869. fol.

On African Tragacanth. *Pharm. Journ.* II. x. (1869) 641-643.

Rhatany from Para. *Pharm. Journ.* III. i. (1870) 84-86.

Nigella seeds or Black Cummin. *Pharm. Journ.* III. ii. (1871)
161.

[Thomas A. C.]
A manual of gardening for Bengal
& Upper India
8v. W. Calcutta. 1874. 8°.

FLUECKIGER (Friedrich A.) *continued :*—
Wild rue or Harmal seeds (Semen Harmalae). *Pharm. Journ.*
III. ii. (1871) 229.

The mother plant of Wormseed. [Artemisia Cina, *Willk.*] Abstracted from a paper of Professor Willkomm. *Pharm. Journ.*
III. ii. (1872) 762–763.

On the occurrence of Manganese in plants, especially in drugs of the Zingiberaceous order. *Pharm. Journ.* III. iii. (1872) 208–211.

Grundlagen der pharmaceutischen Waarenkunde. Einleitung in das Studium der Pharmacognosie. Berlin, 1873. 8°. Ed. 2. 1880.

Buchu leaves. [Barosma betulina. Extract.] *Pharm. Journ.*
III. iv. (1845) 689–690.

Note upon Rhatany from Ceará. [Extract.] *Pharm. Journ.*
III. vi. (1875) 21.

Note on Hing of the Bombay market, the so-called nauseous Asafoetida. *Pharm. Journ.* III. vi. (1875) 401.

Remarks upon Rhubarb and Rheum officinale. *Pharm. Journ.*
III. vi. (1876) 861–863.

Contributions towards the history of some drugs. [Abstract.]
Pharm. Journ. III. vi. (1876) 1021–1023.

Note on a so-called Wood oil. [Extract.] *Pharm. Journ.* III.
vii. (1876) 2–3.

Note on Costus. *Pharm. Journ.* III. viii. (1879) 121.

Note on Luban-mati and Olibanum. *Pharm. Journ.* III. viii.
(1878) 805–908.

Notes on Chian Turpentine. *Pharm. Journ.* III. xi. (1880) 309.

Notes on Cananga oil, or Ilang-ilang oil. [*Archiv der Pharm.*]
Pharm. Journ. III. xi. (1881) 934–937.

FLUECKIGER (Friedrich A.), & E. BURI.
Contribution to the history of Kosin. [Hagenia abyssinica, *Lam.*
Extract.] *Pharm. Journ.* III. v. (1875) 562–563.

FLUECKIGER (Friedrich A.), & Daniel HANBURY.
Pharmacographia : a history of the principal Drugs of vegetable origin met with in Great Britain and British India. London, 1874. 8°. Ed. 2. 1880.

—— Histoire des drugs d'origine végétale . . . Tr. par J. L. de Lanessan, avec une préface par H. Baillon et 320 figures . . . par L. Hugon. Paris, 1878. 2 vols. 8°.

FLUECKIGER (Friedrich A.), & —. HOERN.
Ophelia Chirayta. *Pharm. Journ.* III. i. (1870) 105–107.

FORBES (John).

Salicetum Woburnense: or a catalogue of willows indigenous or foreign in the collection of the Duke of Bedford, at Woburn Abbey; systematically arranged. [London,] 1829. 4°. [*Anon.*]

Pinetum Woburnense: or a catalogue of coniferous plants, in the collection of the Duke of Bedford, at Woburn Abbey, systematically arranged. Londini, 1839. 8°. [*Anon.*]

FOREST (H.).

Du cacao, et de ses diverses espèces. Histoire, analyses chimiques, alterations et falsifications, *etc.* Paris, 1864. 8°.

FORSTER (Johann Georg Adam).

Geschichte und Beschreibung des Brodbaums. Programm. Cassel, 1784. 8°.

De plantis esculentis insularum oceani australis commentatio botanica. Berolini, 1786. 8°.

FORTUNE (Robert).

Three years' wanderings in the Northern Provinces of China, including a visit to the Tea, Silk, and Cotton countries, with an account of the agriculture and horticulture of the Chinese, new plants, *etc.* London, 1847. 8°.

The Tallow Tree. [From Fortune's Wanderings.] *Pharm. Journ.* vii. (1847) 289–290.

Report upon the Tea plantations in the North-Western Provinces, *etc.* [London, 1851.] 8°.

A journey to the Tea countries of China, including Sung-lo and the Bohea Hills, with a short notice of the East India Company's Tea plantations in the Himalaya mountains. London, 1852. 8°.

Two visits to the tea countries of China and the British plantations in the Himalayas, with a narrative of adventures, and a full description of the culture of the tea plant, *etc.* London, 1853. 2 vols. 8°.

A residence among the Chinese, *etc.* London, 1857. 8°.

Yedo and Peking. A narrative of a journey to the capitals of Japan and China, with notices of the natural productions, agriculture, *etc.* London, 1863. 8°.

FORSYTH (Alexander).

Medicinal herbs and quackery. [*Gardeners' Chronicle.*] *Pharm. Journ.* III. ii. (1871) 507–508.

FOURCADE (Charles).

Herbier agricole pour faciliter la connaissance et la culture des plantes utiles. Luchon, 1874. 4°.

FOURNIER (E. H.).

On Coca. [*Food Journal.*] *Pharm. Journ.* III. i. (1870) 43–44.

FOURNIER (Eugène).

Des ténifuges employés en Abyssinie. Paris, 1861. 4°.

FOWKE (R.).

Experiments on the strength of British, Colonial, and other Woods. London, 1867. 8°. *

FRAAS (Carl).

Beitrag zur Geschichte europaeischer Kulturpflanzen. Programm der Gewerbschule zu Freysing. Freysing, 1843. 4°.

FRAGOSO (Juan).

Discursos de las cosas aromaticas, arboles y frutales y de otras muchas medicinas simples, que se traen de la India oriental y sirven al uso de medicinia. Madrid, 1572. 8°.

———— Aromatum, fructuum et simplicium aliquot medica-mentorum, *etc.* Argentinae, 1601. 8°.

FRANCKE (L.).

Wand-Bilder-Atlas der wichtigsten Arznei- und Giftpflanzen Mittel-Europas. Systematisch zusammengestellt. Mit text. Nuernberg, 1863. fol.

FRANÇOIS DE NEUFCHATEAU (Nicholas Louis).

Supplément au mémoire de M. Parmentier sur le mais (ou plutôt maïz). Paris, 1817. 8°.

FRANK (A.), & J. GRUBER.

Tabelle zur Bestimmung der in Deutschland wildwachsenden Holzgewaechse (Baeume und Straeucher) fuer angeh. Botaniker, Forst-Eleven, Lehrer, Touristen, *etc.* Wien, 1879. 8°.

FRANK (B.).

Dei Krankheiten der Pflanzen. Breslau, 1880. 8°.

FRASER (Horatio W.).

Cotton Seeds. [*Amer. Journ. Pharm.*] *Pharm. Journ.* III. ii. (1872) 867–868.

FRASER (Thomas R.).

The Kombé arrow poison (Strophanthus) of Africa. [Abstract.] *Pharm. Journ.* III. iii. (1873) 523–524.

FRÉMY (Edmond), & Pierre Paul DEHÉRAIN.

Recherches sur les betteraves à sucre. Paris, 1875. 8°.

FRIEBE (Wilhelm Christian).

Oekonomisch-technische Flora fuer Liefland, Esthland und Kurland. Riga, 1805. 8°.

FRIES (Elias Magnus).

Sveriges aetliga och giftige svampar, tecknade efter naturen, *etc.* Stockholm, 1862–69. fol.

FRIES (Ludwig von, & Sigmund von).

Uebersichtliche Darstellung der Thee-Cultur und des Thee-Handels in China, als MS. gedruckt. Wien, 1878. 8°.

FRONMUELLER (——).

On Indian Hemp. [Extract.] *Pharm. Journ.* II. ii. (1860) 225–227.

FRUEHLING (R.), & J. SCHULZ.

Anleitung zur Untersuchung der fuer die Zucker-Industrie in Betracht kommenden Rohmaterialen, Producte, Nebenproducte, *etc.* Braunschweig, [1876]. 8°.

FUCHS (Joseph).

Katalog der Hoelzer-Sammlung des allgemeinen oesterreichischen Apotheker-Vereines. Wien, 1866. 8°.

FUENTES (Manuel A.).

Mémoire sur le coca de Pèrou, ses caractères botaniques, sa culture, ses propriétés hygieniques et thérapeutiques. Paris, 1866. 8°.

FYFE (Alexander Gordon).

Suggestions for separating the Culture of Sugar from the process of Manufacture, *etc.* London, 1846. 8°.

GALL (Ludwig).

Anleitung fuer den Landmann zur Syrup- und Zuckerbereitung aus Kartoffeln mittelst gewoehnlichter Branntweinbrennerei-Geraethe, *etc.* Trier, 1825. 8°.

GALLAIS (—.).

Cultivation of Rhubarb in France. [*Bull. Soc. Acclimatation.*] *Pharm Journ.* III. xi. (1881) 755–756.

GALLESIO (Georgio).

Traité du Citrus. Paris, 1871. 8°.

GAMBLE (J. Sykes).

On the State Forests and Forest Schools of France. Edinburgh, 1872. 8°.

List of the Trees, Shrubs, and large Climbers found in the Darjeeling district, Bengal. Calcutta, 1878. 8°.

GARNERONE (G.).

Cenni sul sorgo zuccherino, sua vera coltivazione come piante do zucchero e da foraggio. Torino, 1873. 16°.

GARREAU (—.), & —. MACHELART.

Nouvelles recherches sur les saxifrages, applications de leurs produits aux arts et à la therapeutique, experience sur leur culture. Bailleul, 1881. 8°.

GARRIGUES (—.).

Culture du bambou dans les Basses-Pyrénées. Paris, 1879. 8°.

GASTINEL (—.).

Monographie des Opiums de la haute Égypte. [*Mem. de l'Inst. égyptien*, tome i.] Paris, 1862. 4°.

GAY (Jacques).

Le Chamaerops excelsa (Thunb.) . . . son fibrillitium, les usages économiques auxquels il peut servir, *etc. Bull. Soc. bot. de France*, viii. (1861) 410–430.

GAYER (Karl).

Der Waldbau. Berlin, 18 –80. 2 vols. 8°.

GAYFFIER (Eugène de).

Herbier forestier de la France. Reproduction par la photographie d'après nature et de grandeur naturelle de toutes les plantes ligneuses qui croissent spontanément en forêt. Description botanique, situation, culture, qualité, usages, *etc.* Paris, 1868–73. 2 vols. fol.

GERMAIN (F.).

Rapport sur le reboisement, présenté au conseil général de la Drôme (Août, 1873). Valence, 1875. 8°.

GENTIL (André Antoine Pierre).

Dissertation sur le café, et sur les moyens propres à prevenir les effets qui résultent de sa preparation, communément vicieuse, et en rendre la boisson plus agréable et plus salutaire. Paris, 1797. 8°.

GENTILE (G.).

Monografia sulle piante forestali, industriali e fruttifere, spontanee o naturalizzate, nel circondario di Porto Maurizio, formanti la collezione xilologica presentata da comizio agrario al concorso regionale di Genova. Oneglia, 1879. 8°.

GENTH (Gustav).

Doppelte Riefen. Eine neue Methode zur Erziehung des Laubholzes fuer Waldeigenthuemer und Forstkundige. Trier, 1874. 16°.

GERRARD (A. W.).

" Wanika," a new African arrow poison: its composition and properties [Strophanthus]. *Pharm. Journ.* III. xi. (1881) 833–834. *See also* pp. 849–850.

GERWIG (Friedrich).

Die Weisstanne (Abies pectinata, DC.) im Schwarzwalde. Ein Beitrag zur Kenntniss ihrer Verbreitung, ihres forstlichen Verhaltens und Werthes, ihrer Behandlung und Erziehung. Berlin, 1868. 8°.

GEYER (C.).

Anbau und Pflege derjenigen fremdlaendischen Laub- und Nadelhoelzer, welche die norddeutschen Winter erfahrungsmaess im Freien aushalten, *etc.* Berlin, 1872. 8°.

GEYLER (H. Th.).

Ueber Culturversuche mit dem Japanischen Lackbaum, Rhus vernicifera, DC., im botan. Garten zu Frankfurt-am-Main. Frankfurt-am-Main, 1881. 4°.

GIANFILIPPI (Filippo A. de').

Istruzione popolare sulla coltivazione dell' ulivo. Bardolino, 1874. 16°.

GIBB (George D.).

Examination of the Sap of the Sugar Maple tree, the Acer saccharinum of Linnaeus, with an account of the preparation of the sugar. [From the *Brit. Amer. Journ. Med. Science.*] *Pharm. Journ.* xi. (1859) 115–119.

The description, composition, and preparations of the Sanguinaria canadensis. *Pharm. Journ.* II. i. (1860) 454–463.

GIBBS (Joseph).

Cotton Cultivation in its various details . . . especially adapted to the improvements of the cultural soils of India. London, 1862. 8°.

GIBBS (W.).

New Zealand Flax, or Phormium tenax. An account of the plant and its use for Manufacturing purposes. Printed on Paper made from its Fibre at Wraysbury Mill, 1864. [London, 1865 ?] 8°.

GIBBS (W. A.).

On the Cultivation of Beetroot, and its Manufacture into Sugar. *Journ. Soc. Arts*, xvi. (1868) 415–424.

GIBELLI (G.).

La malettia del castagno, osservazioni ed esperienze. Modena, 1880. 8°.

GIBSON (Alexander).

Cortex Alstoniae scholaris. *Pharm. Journ.* xii. (1853) 422–423.

On Kino. *Pharm. Journ.* xii. (1853) 496–497.

GIERSBERG (Franz, & J. W.).

Krankheiten der landwirthschaftlichen Culturpflanzen. (4 Hefte.) Leipzig, 1878. 8°.

GILDAS (—.).

L'eucalyptus dans la campagne romaine. Lettre adressé à M. le président de la société d'acclimatation. Paris, 1875. 8°.

GILES (R. W.).
> The official, commercial and authentic sources of Pareira Brava.
> *Pharm. Journ.* III. iv. (1873) 519–520.

GILL (C. Haughton).
> On the Manufacture and Refining of Sugar. *Journ. Soc. Arts,*
> xx. (1871) 111–115, (1872) 128–131, 140–144, 162–165.

GILLET-DAMITTE (—.).
> Le galéga, nouveau fourrage, sa culture, son usage et son profit.
> Paris, 1867. 18°.

GILMOUR (William).
> Gurgun Balsam. *Pharm. Journ.* III. v. (1875) 729–730.

GIMBERT (—.).
> L'Eucalyptus Globulus, son importance en agriculture, en hygiène
> et en médicine. Paris, 1870. 8°.
> —— El Eucalyptus Globulus, su importanica en agricultura, en
> higiene y en medicina. Clichy, 1873. 8°.

GLASSPOOLE (Hampden Gledstanes).
> Cork, corks and corkscrews. *Pharm. Journ.* III. ix. (1879)
> 995–997.

GLEDITSCH (Jchann Gottlieb).
> Alphabetisches Verzeichniss der gewoenlichsten Arzneigewaechse,
> ihrer Theile und rohen Produkte, welche in den groessten
> deutschen Apotheken gefunden werden. Berlin, 1769. 8°.
> Systematische Einleitung in die neue Forstwissenschaft. Berlin,
> 1775. 2 vols. 8°.
> Einleitung in die Wissenschaft der rohen und einfachen Arznei-
> mittel. Berlin und Leipzig, 1778–87. 4 vols. 8°.
> Botanica medica, oder die Lehre von den vorzueglich wirksamen
> einheimischen Arzneigewaechsen. Herausgegeben von F. W. A.
> Lueders. Berlin, 1788–89. 2 vols. 8°.

GLOVER (R. D.).
> Antimalarial properties of the Eucalyptus. *Pharm. Journ.* III.
> vi. (1876) 625. *See also* p. 639.

GMELIN (Johann Friedrich).
> Abhandlung von den giftigen Gewaechsen, welche in Teutschland
> und vornehmlich in Schwaben wild wachsen. Ulm, 1775. 8°.
> Also dated 1805.
> Allgemeine Geschichte der Pflanzengifte. Nuernberg, 1777. 8°.
> Ed. 2. 1803.

GOBIN (A.).
> Guide pratique pour la culture des plantes fourragères. Paris,
> 1865. 18°.

GOCHNAT (Franz Karl).

Tentamen medico-botanicum de Cichoraceis.　Argentorati, 1808.
4°.

GODRON (Dominique Alexandre).

Les cuscutes et leurs ravages dans nos cultures. Nancy, [1876].　8°.

GOEBEL (Carl Christian Traugott Friedmann), & Gustav KUNZE.

Pharmaceutische Waarenkunde, mit illuminirten Kupfern nach
der Natur gezeichnet von Ernst Schenk.　Eisenach, 1827–34.
2 vols.　4°.

GOEPPERT (Heinrich Robert).

Ueber die giftige Pflanzen Schlesiens.　Programm.　Breslau,
[1832].　8°.

Die in Schlesien wildwachsenden offizinellen Pflanzen.　Ein-
ladungs Programm.　Breslau, 1835.　8°.

Die officinellen Gewaechse Europaischer botanischer Gaerten,
insbesondere die des Koenigl. botanischen Gartens der
Universitaet Breslau.　(Aus dem *Archiv der Pharmacie*, 1863,
abgedruckt).　Hannover, 1863.　8°.

GOFFART (Auguste).

Sur la culture et l'ensilage du maïs-fourrage.　Mémoire présenté
à la société centrale d'agriculture de France.　[2 parts.]
Paris, 1875.　8°.

Manuel de la culture et de l'ensilage des maïs et autres fourrages
verts.　3e édition.　Paris, 1879.　8°.

GOMES (Bernardino Antonio).

Observationes botanico-medicae de nonnullis Brasiliae plantae,
. . . (Observações botanico-medicas sobre algumas plantas do
Brazil, etc.).　Olissipone, 1803.　(2 fasc.)　4°.

GOMICOURT (— de).

De l'ailanthe globuleux ou vernis de Japon et de son utilité.
[From the *Revue agricole du Gers.*]　Auch, 1870.　8°.

GONCET DE MAS (—.).

Culture de la ramie.　[Extract.]　Paris, 1877.　8°.

GORDON (C. A.).

Influence of vegetation upon Malaria.　[Extract.]　*Pharm. Journ.*
III: iv. (1874) 777–778.

GORDON (George), & Robert GLENDINNING.

The Pinetum ; being a synopsis of all the coniferous plants at
present known, with descriptions, history and synonymes, and
comprising nearly one hundred new kinds.　London, 1858.　8°.
Supplement, 1862.　Ed. 2. with Index by H. G. Bohn, 1875.
Ed. 3. 1880.

yer, C., Gum-elastic and its varieties, with a detailed ac
applications and the discovery of vulcanisati
8°. NewHaven, Conn., 1855.

GORINI (G.).
  Piante industriale : descrizione, coltivazione, raccolta e prepara-
    zione. Milano, 1878. 32°.
GORKOM (K. W. van).
  Die Chinacultur auf Java. Aus dem Hollandischen uebertragen
    von C. Hasskarl. Leipzig, 1869. 8°.
  The cultivation of Cinchonas in Java. London, 1870. 8°.
  Cinchona cultivation in Java. [Extract from letter.] *Pharm.
    Journ.* III. iii. (1872) 31.
  Cinchona cultivation in Java. [*Pharm. Zeitung.*] *Pharm. Journ.*
    III. iv. (1874) 656–657.
  Wetenschappelijke opmerkingen en ervaringen betreffende de
    Kinakultuur. *Verslagen Kon. acad. wetenschappen,* II. xiv.
    (1879).
  De Oost-Indische cultures in betrekking tot handel en nijverheid.
    Vol. i. Amsterdam, 1881. 8°.
GOSSE (L. A.).
  Monographie de l'Erythroxylum Coca. Bruxelles, 1861. 8°.
GOURDON (Jean), & P. NAUDIN.
  Nouvelle iconographie fourragère : comprenant un atlas, avec text
    explicatif des plantes fourragères et des plantes nuisibles qui
    se rencontrent dans les prairies et les pâturages, *etc.* Paris,
    1865–71. 8°.
GRAF (Sigmund).
  Die Fieberrinden in botanischer, chemischer, und pharma-
    ceutischer Beziehung dargestellt. Wien, 1824. 8°.
GRANDEAU (L.).
  Cours d'agriculture de l'école forestière. Chimie et physiologie
    appliquées à l'agriculture et à la sylviculture. I. La nutrition
    de la plante ; l'atmosphère et la plante. Paris, 1879. 8°.
GRAUMUELLER (Johann Christian Friedrich).
  Flora pharmaceutica Jenensis. Jena, 1815. 4°.
  Handbuch der pharmaceutisch-medizinischen Botanik. Eisenberg,
    1813–19. 6 vols. 8°.
GRAVES (George), & J. D. MORRIES.
  Hortus medicus. Edinburgh, 1834. 8°.
GREBE (Carl).
  Der Waldschutz und der Waldpflege. [Ed. 3. of G. Koenig's
    Waldpflege.] Gotha, 1875. 8°.
GREENE (Edward Lee).
  Notes on certain silkweeds. [Asclepias Meadii, *Torr.,* A. obtusi-
    folia, *Mich.,* A. Sullivantii, *Engelm.,* A. speciosa, *Torr.,* A.

GREENE (Edward Lee) *continued :*—

Cornuti, *Dcne.*, and A. uncialia, *Greene.*] *Coulter's Bot. Gaz.*
v. (1880) 64–5.

GREENISH (Henry George).

Cape tea. *Pharm. Journ.* III. xi. (1881) 549–551, 569–573.

GREENISH (Thomas).

The Flax lints of commerce under the microscope. *Pharm. Journ.*
III. i. (1870) 352–354.

Linseed and Linseed Meal. *Pharm. Journ.* III. ii. (1871)
211–213.

Note on Scammony. *Pharm. Journ.* III. v. (1874) 263–264. *See
also* 191–193.

The Microscopy of Natal Arrowroot. *Pharm. Journ.* III. vi. (1875)
204–208.

Arrowroot. *Pharm. Journ.* III. vii. (1876) 169–170.

Further researches on Tea Hair. *Pharm. Journ.* III. viii. (1877)
250.

The microscope in Materia medica. *Pharm. Journ.* III. ix. (1878)
193–195.

The histology of Araroba or Goa powder. *Pharm. Journ.* III. x.
(1880) 814–816.

GRÉGOIRE (J.).

De la culture du coton en Egypte. Historique. État actuel.
Avenir. [*Mémoires de l'Institut égyptien, Tome i.*] Paris,
1862. 4°.

GRESSLER (Friedrich Gustav Ludwig).

Deutschlands Giftpflanzen mit den naturgetreuen Abbildungen,
aus der Naturgeschichte. Langensalza, 1853. 8°. Ed. 6.
1864. Ed. 7. 1868. Ed. 8. Stettin, 1872. Ed. 10. Langensalza, 1876.

GREY (George).

Herb-poisoning at the Cape of Good Hope. [*Brit. Med. Journ.*]
Pharm. Journ. III. v. (1874) 248–250.

GRIFFEN (Augustus R.).

An essay on the botanical, chemical, and medical properties of
the Fucus edulis of Linnaeus, *etc.* New York, 1816. 8°.

GRIFFITH (William).

Report on the Tea Plant of Upper Assam. Calcutta, 1838. 8°.

GRIGOLATO (Gaetano).

Flora medica del Polesine, ovvero descrizione delle piante
medicinali che nascono nelle provincia di Rovigo. (Fasc.
1–5.) Rovigo, 1843. 4°.

GRIMALDI (Luigi).

Sulla coltivazione ed industria del cotone in Europa e precipuamente nella Calabria ultra seconda. Torino, 1863. 8°.

GRIMAUX (Edouard).

Du hachisch ou chanvre indien. Paris, 1865. 8°.

GRINDEL (David Hieronymus).

Pharmaceutische Botanik zum Selbstunterrichte, insbesondre fuer angehende Apotheker und Aerzte. Riga, 1862. 8°.

GRINDON (Leopold Hartley).

Trees of Old England; sketches of the aspects, associations, and uses of those which constitute the forests, and give effect to the scenery of our native country. London, 1868. 8°. Ed. 2. 1872.

GRIPOUILLEAU (Armel).

Le mélilot jaune ou mélilot officinal, le mélilot blanc ou mélilot de Sibérie, plantes fourragères très-mellifères. Tours, 1864. 4°.

GROENLUND (Chr.).

Danske Giftplanter. Kjoebenhavn, 1874. 8°.

GROSOURDY (Réné de).

El médico botánico criollo. Paris, [1864]. 4 vols. 8°.

GROSS (G.).

Die wichtigern Handelspflanzen in Bild und Wort. Esslingen, 1880. fol.

GROSSE (Ernst).

Deutschlands Kulturpflanzen. Leipzig, 1858. 8°. Ed. "2." 1862.

GROTHE (Hermann).

Die Gespinnstfasern aus dem Pflanzenreich. Nach den Materialen der Austellungen in London, Petersburg, Neapel, Kopenhagen, Amsterdam, Moskau, Mailand, Wien, Philadelphia, Paris, und A. Berlin, 1879. 8°.

———— Notes on some neglected Fibres. Edited by C. G. W. Lock. *Journ. Soc. Arts,* xxviii. (1880) 912–914, 916–920.

Gespinnstfasern aus Agaven. (*Dingler's . . . Zeitung* (1880) 157.)

GROVES (Henry).

Note on the cultivation and preparation of Castor-oil in Italy. *Pharm. Journ.* II. viii. (1866) 250–252.

Florentine Orris. *Pharm. Journ.* III. iii. (1872) 229–231.

Note on some indigenous Tuscan remedies. *Pharm. Journ.* III. v. (1874) 230–233.

GROVES (T. B.).

On Portland Arrow-root. *Pharm. Journ.* xiii. (1853) 60–61.

Nepaul aconite. *Pharm. Journ.* III. i. (1870) 433–434.

Note on Flower of tea, or Pekoe Flower. *Pharm. Journ.* III. vii. (1876) 285–286.

GUBLER (Adolphe).

The Eucalyptus Globulus and its use in medicine. [*Journ. de Pharm.*] *Pharm. Journ.* III. ii. (1872) 703–704.

Note on a Piper, called Jaborandi, in the province of Rio Janeiro. [Extract.] *Pharm. Journ.* III. vii. (1877) 731–732.

GUENTHER (Johann), & Friedrich BERTUCH.

Pinakothek der deutschen Giftgewaechse. Jena, 1840. 4°.

GUIBOURT (Nicolas Jean Baptiste Gaston).

Histoire abrégée des drogues simples. Paris, 1822. 2 vols. 8°. Ed. 2. 1826. Ed. 3. 1836. Ed. 4. 1849–51, 4 vols. Ed. 5. [no alterations] 1857. Ed. 6. (by G. Planchon) 1868–70. Ed. 7. 1876.

On a false Jalap having a rose-odour. *Pharm. Journ.* ii. (1842) 331–337.

On the galls of Terebinthus and Pistacia. *Pharm. Journ.* iii. (1844) 377–380.

———— Appendix to the preceding paper. *op. cit.* 381–383.

Mémoire sur les sucs astringents connues sous les noms de cachou, gambir et kino. Paris, [1847]. 8°.

Mémoire sur le tabaschir. Paris, 1855. 8°.

On Tragacanth and some allied Gums. Translated from a letter addressed to Mr. Daniel Hanbury. *Pharm. Journ.* xv. (1855) 57–59.

Note on Wood Oil. [From the *Journ. de Pharm.*] *Pharm. Journ.* xvi. (1856) 332–334.

GUILLEMEAU (Jean Louis Marie).

Histoire naturelle de la rose, où l'on décrit ses différentes espèces, sa culture, ses vertus et ses propriétés, etc. Paris, 1800. 12°.

GUILLEMIN (Jean Antoine).

Rapport au ministre de l'agriculture, sur sa mission au Brésil, ayant pour objet des récherches sur les cultures et la préparation du thé, et le transport de cet arbuste en France. Paris, 1839. 8°.*

GUIMPEL (Friedrich).

Abbildung und Beschreibung aller in der Pharmacopoeia borussica aufgefuehrten Gewaechse. Text von D. F. L. Schlechtendal. Berlin, 1830–37. 3 vols. 4°.

GUIMPEL (Friedrich) *continued*:—
Pflanzenabbildungen und Beschreibungen zur Erkenntniss offizineller Gewaechse. Text von Johann Friedrich Klotsch. Berlin, 1838. 4°.

GUIMPEL (Friedrich), Carl Ludwig WILLDENOW, & Friedrich Gottlob HAYNE.
Abbildung der deutschen Holzarten. Berlin, 1815-20. 2 vols. 4°.

HABERLANDT (Friedrich).
Die Sojabohne. Ergebnisse der Studien und Versuche ueber die Anbauwuerdigkeit dieser neu einzufuehrenden Culturpflanzen. Wien, 1878. 8°.

HAERING (—.).
Zusammenstellung der Kennzeichen der in Deutschland wachsenden verschiedenen Eichen-Gattungen und ihrer hauptsaechlichen Fehler; insbesondere zum Anhalt bei der Abnahme von Eichenhoelzern fuer die Marine. Berlin, 1853. fol.

HAGENDORP (C. S. W., *Comte de*).
Coup-d'œil sur l'ile de Java et les autres possessions néerlandaises dans l'archipel des Indes. Bruxelles, 1830. 8°.

HAHMANN (A.).
Die Dattelpalme, ihre Namen und ihre Verehrung in der alten Welt. Nordhausen, 1858. 8°.

HAINES (R.).
Notes on Conessine, alias Wrightine. [*See also* STENHOUSE.] *Pharm. Journ.* II. vi. (1865) 432–434.

HALL (Hermann Christian van).
Toegepaste kruidkunde. Handleiding tot de aanwijzing van het gebruik, dat de mensch maakt van voorwerpen uit het plantenrijk. Groningen, 1857. 8°.

HALLE (Johann Samuel).
Die deutsche Giftpflanzen. Berlin, 1784-93. 2 vols. 8°.

HALLEZ D'ARROS (—.).
De l'avenir de la culture et de l'industrie du lin en Algérie. Paris, 1880. 8°.

HAMBURGER (Z. S.).
Ueber die Farbstoffe der Quercitronrinde. Goettingen, 1880. 8°.

HANAUSEK (T. F.).
Mittheilungen aus dem Laboratorium der Waarensammlung in Krems. Die Tahitinuss. [Sagus amicarum, *Wendl.*] (*Zeitschr. allg. oest. Apoth. Ver.* 1880, No. 28.)

HAMILTON (William).

On the Jamaica Dogwood. *Pharm. Journ.* iv. (1844) 76–78.

Further remarks on Piscidia erythrina, or Jamaica Dogwood. *Pharm. Journ.* iv. (1844) 111–114.

On the Medicinal properties of the Argemone mexicana and Hura crepitans. *Pharm. Journ.* iv. (1844) 167–170.

On the gum of the Rhus metopium, and on the Aristolochia odoratissima, [A.] trilobata, and [A.] anguicida. *Pharm. Journ.* v. (1844) 60–61.

On the Argemone mexicana, Hura crepitans, Jatropha Curcos, and [J.] multifida. *Pharm. Journ.* v. (1845) 23–27.

On the genus Jatropha, and the Janipha Mainhot and [J.] Loeflingii. *Pharm. Journ.* v. (1845) 27–31.

—————— Sequel to the above paper on Janipha. *op. cit.* 31–33.

Medical properties of the Fevillea cordifolia. *Pharm. Journ.* v. (1845) 33–35.

On the Moringa pterygosperma, or Oil of Ben tree, and its uses economical and officinal. *Pharm. Journ.* v. (1845) 58–59.

On the properties ascribed to some species of Comocladia. *Pharm. Journ.* v. (1845) 114–117.

On the properties of some species of Cathartocarpus. *Pharm. Journ.* v. (1845) 118–121.

On the medical and economical properties of the Anacardium occidentale, or Cashew-nut tree. *Pharm. Journ.* v. (1845) 268–273.

On the medical properties of the various species of Cassia, found in the West Indies. *Pharm. Journ.* v. (1846) 345.

On the properties of the Hippomane Mancinella, or Manchineel tree. *Pharm. Journ.* v. (1846) 408–412.

On the medical and commercial properties of the Caesalpinia coriaria, or Dividivi. *Pharm. Journ.* v. (1846) 443.

On the noxious properties of the Echites suberecta, or Savanna flower, and their antidotes. *Pharm. Journ.* vi. (1846) 23–24.

On the Maranta arundinacea, and its application as an antidote to animal and vegetable poisons. *Pharm. Journ.* vi. (1846) 25–28.

On the reputed properties of some of the species of Cleome. *Pharm. Journ.* vi. (1846) 79–80.

On the Capparis cynophallophora, or Bottle cod. *Pharm. Journ.* vi. (1846) 81–82.

On the properties of the Asclepias curassavica, a bastard Ipecacuanha *Pharm. Journ.* vi. (1846) 212–216.

HAMILTON (William) *continued* :—

On the Muracuja ocellata, or West Indian opium. *Pharm. Journ.* vi. (1846) 216–217.

On the Malambo bark of the province of Carthagena, South America. *Pharm. Journ.* vi. (1846) 255–259.

On some species of the genus Amyris, which are little known. *Pharm. Journ.* vi. (1847) 322–326.

On the Lemon Grass, or Andropogon Schoenanthus, cultivated in most of the West Indian Islands. *Pharm. Journ.* vi. (1847) 369–374.

On the commercial and medicinal properties of the Hymenaea Courbaril, or Locust tree, etc. *Pharm. Journ.* vi. (1847) 520– [525].

On the properties of the Bocconia frutescens and Clusia. *Pharm. Journ.* vi. (1847) 581–582.

On the Canna Achira[s], or Tous les mois. With a specimen of recent tuber. *Pharm. Journ.* vii. (1847) 56–60.

On the medicinal and economic properties of the Sapindus Saponaria, Soap berry, or Black Nickar Tree. *Pharm. Journ.* vii. (1847) 225–230.

Some further particulars respecting the Rhus Metopium, or Hog-gum tree. *Pharm. Journ.* vii. (1847) 270–277. [*Anon.*]

On the purgative properties of the oils obtained from plants indigenous in the Antilles. *Pharm. Journ.* ix. (1849) 129–131.

On the Ava root of the Sandwich Islands. *Pharm. Journ.* ix. (1849) 218–220.

On the cultivation, preparation, and properties of the Myrica cerifera, and M. carolinensis, or Wax-tree of Carolina. *Pharm. Journ.* x. (1851) 450–454.

Use of Mushrooms in Russia. *Pharm. Journ.* xiv. (1854) 66–67.

On the medicinal properties of the Coffee arabica. *Pharm. Journ.* xiii. (1854) 329–330.

On the nutritive properties of the mass which remains after the expression of the nuts of the Cocoa, *etc. Pharm. Journ.* xv. (1856) 350.

Notes on the Materia medica of the Paris Exhibition, 1855. *Pharm. Journ.* xv. (1856) 410–411.

HAMM (Wilhelm von).

Der Fieberheilbaum oder Blaugummibaum (Eucalyptus Globulus). Ueber dessen Anbau und seine Eigenschaft der Gesundmachung von Sumpflaendereien. Nach einem Vortrage von Bentley . . . mit vielen Zusaetzen, *etc.* Wien, 1878. 8°.

HANBURY (Daniel).

On Turnsole. *Pharm. Journ.* ix. (1850) 308–309.

On the Resin of the Norway Spruce Fir (Abies excelsa). *Pharm. Journ.* ix. (1850) 400–401.

On an article imported as Calumba wood, supposed to be the product of a Menispermum. *Pharm. Journ.* x. (1851) 321–323.

On Wurrus, a dye produced by Rottlera tinctoria. *Pharm. Journ.* xii. (1853) 589-590.

On the use of Coffee-leaves in Sumatra. *Pharm. Journ.* xiii. (1853) 207–209.

Notes upon some specimens of Scammony. *Pharm. Journ.* xiii. (1853) 268–270.

On the febrifuge properties of the Olive (Olea europaea, L.). *Pharm. Journ.* xiii. (1854) 353–354.

On some rare kinds of Cardamom. *Pharm. Journ.* xiv. (1855) 352–355, 416-422.

On Wood oil, a substitute for Copaiba. [From Dipterocarpus turbinatus, *Roxb.*] *Pharm. Journ.* xv. (1856) 321–322.

Rhatany Root. *Pharm. Journ.* xvi. (1856) 16–17.

Note upon a green dye from China. *Pharm. Journ.* xvi. (1856) 213–214.

On Penghawar Djambi, a new Styptic. *Pharm. Journ.* xvi. (1856) 278–281.

Some notes on the manufactures of Grasse and Cannes. *Pharm. Journ.* xvii. (1857) 161–163.

On Storax. *Pharm. Journ.* xvi. (1857) 417–423, 461–465.

On Rottlera tinctoria, *Roxb.*, and its medicinal properties. *Pharm. Journ.* xvii. (1858) 405–410.

Note on a drug called Royal Salep. *Pharm. Journ.* xvii. (1858) 499–501.

On Otto of Rose. *Pharm. Journ.* xviii. (1859) 504–510.

Malambo bark [Croton Malambo, *Karst.*] *Pharm. Journ.* II. i. (1859) 321–322.

Note on a manufactured product of Sea-weed called Japanese Isinglass. *Pharm. Journ.* II. i. (1860) 508–509.

Bassia flowers. *Pharm. Journ.* II. i. (1860) 607–608.

Notes on Chinese Materia medica. *Pharm. Journ.* II. ii. (1860) 15–18, 109–116, (1861) 553–557, iii. 6–11, 204–209, 261–264, 315–318, (1862) 420–425.

Note on Anacahuite wood, a reputed remedy for consumption. *Pharm. Journ.* II. ii. (1861) 407–408.

HANBURY (Daniel) *continued :—*

Note on the use of Balsam of Peru in the Roman Catholic Church. *Pharm. Journ.* II. ii. (1861) 446–448.

Notes on Chinese Materia medica. London, 1862. 8°.

> Reprinted as a separate work. Also in German, by T. Martius. Speyer, 1863. 8°.

Minor notes on the Materia medica of the International Exhibition. *Pharm. Journ.* II. iv. (1862) 107–111.

Origin of Anacahuite wood. *Pharm. Journ.* II. iv. (1862), 271–273.

On the manufacture of Balsam of Peru. *Pharm. Journ.* II. v. (1863) 241–248.

Note on Cassia moschata, H. B. K. *Pharm. Journ.* II. v. (1863) 348–351.

Additional observations on Storax. *Pharm. Journ.* II. iv. (1863) 436–439.

Note on the Ordeal bean of Calabar. (Physostigma venenosum, *Balf.*) *Pharm. Journ.* II. iv. (1863) 559–561.

On the species of Garcinia, which affords Gamboge in Siam. *Linn. Soc. Trans.* xxiv. (1864) 487–490.

Additional note on the manufacture of Balsam of Peru. *Pharm. Journ.* II. v. (1864) 315–317.

On the botanical origin of Gamboge. *Pharm. Journ.* II. vi. (1865) 349–350.

On the botanical origin of Savanilla Rhatany. *Pharm. Journ.* II. vi. (1865) 460–461.

On pharmaceutical herbaria. *Pharm. Journ.* II. vii. (1866) 542–544.

On Burgundy Pitch. *Pharm. Journ.* II. ix. (1867) 162–165.

Remarks on the necessity for a further cultivation of medicinal plants. *Pharm. Journ.* II. viii. (1867) 575–577.

On the cultivation of Jalap. *Pharm. Journ.* II. viii. (1867) 651–654.

Historical notes on Manna. *Pharm. Journ.* II. xi. (1869) 326–331.

On a species of Ipomoea (I. simulans), affording Tampico Jalap. [*Linn. Soc. Journ., Bot.*] *Pharm. Journ.* II. xi. (1870) 848–850.

Historical notes on the Radix Galangae of Pharmacy. [*Linn. Soc. Journ., Bot.*] *Pharm. Journ.* III. i. (1871) 613, ii. (1871) 248–249.

The Madagascar Cardamom or Longouze. [Amomum angusti-folium, *Sonn.*] *Pharm. Journ.* III. ii. (1872) 642.

HANBURY (Daniel) *continued* :—

- On Calabrian Manna.   *Pharm. Journ.* III. iii. (1872) 421–422.

African Ammoniacum.   *Pharm. Journ.* III. iii. (1873) 741.

The botanical origin and country of Myrrh.   [Ocean Highways.]
    Pharm. Journ. III. iii. (1873) 821–823.

On Pareira Brava. *Pharm. Journ.* III. iv. (1873) 81–83, 102–103.

On a peculiar camphor from China.   *Pharm. Journ.* III. iv.
    (1874) 709–710.   *See also* p. 722.

Science Papers, chiefly pharmacological . . . edited, with memoir,
    by J. Ince. London, 1876.   8°.

HANBURY (Daniel), & F. A. FLUECKIGER.

Pharmacographia.   *See* FLUECKIGER and HANBURY.

HANBURY (Daniel), & Daniel OLIVER.

Inquiries relating to Pharmacology and Economic Botany. [From
    the *Admiralty Manual of Scientific Enquiry*, Ed. 4.   *See also*
    HERSCHEL (J. F. W.).]   *Pharm. Journ.* III. ii. (1871) 204–205,
    243–245.

HANCE (Henry Fletcher).

The so-called "Olives" (Canarii spp.) of Southern China.
    [*Journ. Bot.*]   *Pharm. Journ.* III. i. (1871) 684–685.

The source of the Radix Galangae minoris of pharmacologists.
    [*Linn. Soc. Journ., Bot.*] *Pharm. Journ.* III. ii. (1871) 246–248.

The Green Putchuk of the Chinese.   With some remarks on the
    antidotal virtues ascribed to Aristolochiae.   [*Journ. Bot.*]
    Pharm. Journ. III. iii. (1873) 725–726.

HANCE (Henry Fletcher), & W. F. MAYERS.

Introduction of Maize into China, Indian Corn, Zea Mays, L.
    With notices of the plant by Chinese authors.   *Pharm.*
    Journ. III. i. (1870) 522–525.

HANCOCK (John).

The properties and preparation of the Rio Negro Sarsaparilla,
    and of the Angustura bark, practically examined.   London,
    1829. 8°.

Remarks on the Siruba, or native Oil of Laurel, its production,
    uses, *etc.* London, 1830. 8°.

Observations on the climate, soil, and productions, of British
    Guiana, *etc.* London, 1835. 8°.   Ed. 2. 1840.

HANCOCK (Thomas).

Personal narrative of the origin and progress of the Caoutchouc
    or India-rubber manufacture in England . . . to which is
    added some account of the plants from which Caoutchouc is
    obtained, *etc.* London, 1857. 8°.

HANSEN (A.)
Die Quebracho - Rinde. Botanisch - pharmaceutische Studie. Berlin, 1880. 4°.

HANSTEIN (Heinrich).
Die Familie der Graeser in ihrer Bedeutung fuer den Wiesenbau. Wiesbaden, 1867. 8°.

HAPPE (Andreas Friedrich).
Abbildungen oekonomischer Pflanzen. Berolini, 1792-94. fol.
Botanica pharmaceutica, exhibens plantas officinales, quarum nomina in dispensatoriis recensentur. Berolini, 1788. fol.

HARDY (Auguste).
Manuel du cultivateur de coton en Algérie. Alger, 1855. 8°. Ed. 2. 1856.
Note sur la culture du Quinquina en Algérie. Paris, 1868. 8°.

HARDY (Ernest).
Pilocarpus pinnatus (Jaborandi) [= P. pennatifolius, *Lemaire*. Extract.] *Pharm. Journ.* III. vii. (1876) 496–498.
Rapport sur l'inée (Strophanthus hispidus) [Extract.] Paris, 1877. 8°.

HARIOT (—.).
Les soixante-quatre plantes utiles aux gens du monde. Troyes, 1876. 8°.

HARMANN (F. E.).
Report on Coffee-leaf Disease (Hemileia vastatrix). Bangalore, 1880. 4°.

HARTIG (Robert).
Die durch Pilze erzeugten Krankheiten der Waldbaeume. Fuer den deutscher Foerster. Ed. 2. Breslau, 1875. 8°.
Die Zersetzungserscheinungen des Holzes der Nadelholzbaeume und der Eiche in forstlicher, botanischer und chemischer Richtung. Berlin, 1878. 4°.
Die Unterscheidungsmerkmale der wichtigeren in Deutschland wachsenden Hoelzer. Specielle Xylotomie. Muenchen, 1879. 8°.
Untersuchungen aus dem forstbotanischen Institut zu Muenchen. Berlin, 1880. 4°.

HARTIG (Theodor).
Lehrbuch der Pflanzenzelle in ihrer Anwendung auf Forstwirthschaft. Berlin, 1841. 4°.
Vollstaendige Naturgeschichte der forstlichen Kulturpflanzen Deutschlands. Berlin, 1851. 4°.
The preceding, with new title; see Pritzel, Thes. ed. 2. No. 3795.

HARTIG (Theodor) *continued :—*

System und Anleitung zum Studium der Forstwissenschaftslehre. Leipzig, 1858. 8°.

Ueber den Gerbstoff der Eiche. Fuer Lederfabricanten, Wald- besitzer · und Pflanzenphysiologen bearbeitet. Stuttgart, 1869. 4°.

HARTINGER (Anton).

Deutschlands Forstculturpflanzen in getreuen Abbildungen . . . Text von Ferdinand Fiscali, *etc.* Ollmuetz, 1854–58. fol.

Die essbaren und giftigen Pilze in ihren wichtigsten Formen. Wien, 1858. fol. Ed. 2. 1870.

HARTUNG (C. A. F. A. Heinrich).

De Cinchonae speciebus atque medicamentis chinam supplentibus. Argentorati, 1812. 4°.

HARTWIG (J.), & Th. RUEMPLER.

Illustrirtes Gehoelzbuch. Die schoensten Arten der in Deutsch- land . . . Baeume und Straeucher, *etc.* Berlin, 1875. 8°.

HARZ (C. O.).

Opium production in Europe. [*Zeitschr. Apoth. Vereins.*] *Pharm. Journ.* III. ii. (1871) 223–225. ,

HARZER (Karl August Friedrich).

Naturgetreue Abbildungen der vorzueglichsten essbaren, giftigen und verdaechtigen Pilze, *etc.* Dresden, 1842–45. 4°.

HASSALL (Arthur Hill).

Food and its adulterations. [From *The Lancet.*] London, 1855. 8°.

On the adulteration of Annatto. [Bixa orellana, *Linn.*] *Pharm. Journ.* **xv.** (1856) "592" [295]–303.

Adulterations detected ; or, plain instructions for the discovery of frauds in Food and Medicine. London, 1857. 8°. Ed. 2. 1861.

Food : its adulterations and the methods for their detection. London, 1876. 8°.

This is not the same as the preceding work.

HASSELT (A. W. M. van).

Bijdrage tot de kennis van het Curare ; met één naschrift van C. A. J. A. Oudemans. *Verslagen kon. acad. weten.* II. **xv.** (1880) pp. 1–12.

HASSKARL (Justus Carl).

Aanteckeningen over het nut, door de bewoners von Java aan eenige planten van het eiland toegeschreven, uit berigten der inlanders zamengesteld. Amsterdam, 1845. 8°.

HAVENSTEIN (G.).

Beitraege zur Kenntniss der Leiupflanze und ihrer Cultur. Physiologisch und landwirthschaftlich begruendet. Inaugural Dissertation. Goettingen, 1874. 8°.

HAWKES (Henry Philip).

Madras Exhibition of 1855. Catalogue raisonné of the thirty classes into which the articles in the Exhibition are divided, with an index of the subjects, *etc.* Madras, 1855. fol.

Report upon the Oils of Southern India. [Madras, 1855.] 8°.

A classified catalogue of the Raw Produce, exhibited at the Madras Exhibition of 1857, *etc.* Madras, 1857. 4°.

HAY (David).

The Pine-tree in New Zealand, being a list of pines . . . with a few hints to those persons who may wish to cultivate them for ornament, for shelter, or for their commercial value. Auckland, 1869. 8°.

HAY (Drummond).

The Argan tree. [Argania Sideroxylon. Extract from Consular report.] *Pharm. Journ.* III. ix. (1878) 262.

HAYNE (Friedrich Gottlob).

Dendrologische Flora der Umgegend und der Gaerten Berlin's. Berlin, 1822. 8°.

Getreue Darstellung und Beschreibung der in der Arzneikunde gebraeuchlichen Gewaechse, wie auch solcher, welche mit ihnen verwechselt werden koennen. Berlin, 1805–46. 14 vols. 4°.

> Vols. 12-14 were posthumous, and brought out by J. F. Brandt, J. T. C. Ratzeburg, and J. H. Klotsch. There is a spurious impression, Berlin, 1829-41, with a slightly altered title, in 4 vols. 4°.

HAYNE (Joseph).

Gemeinnuetziger Unterricht ueber die schaedichen und nuetz-lichen Schwaemme. Wien, 1830. 8°.

HEANEY (John P.).

Megarrhiza californica, *Torrey.* [*American Journ. Pharm.*] *Pharm. Journ.* III. vii. (1876) 393–394.

HEATH (Francis George).

Our woodland trees. London, 1878. 8°.

HECKHELER (Johann).

Dissertatio medica de potù Thèe, *etc.* [M. Mappus, *Praes.*] Argentorati, [1691]. 4°.

HECTOR (James).

Phormium tenax as a fibrous plant. Wellington, 1872. 8°.

HEDGES (Isaac A.).

Sorghum culture and sugar making. *Report, Commissioner of Patents, Agriculture*, 1861 (Washington, 1862) 293–311.

Sorgo, or the northern sugar plant . . . with introduction by W. Clough. Cincinnati, 1863. 12°.

HEFFLER (E. R.).

Notes on the culture of and commerce in opium in Asia minor. [*Extract.*] *Pharm. Journ.* II. x. (1869) 434–437.

HEIN (Heinrich).

Graeserflora von Nord- und Mittel-Deutschland. Eine genaue Beschreibung der Gattungen und Árten der im obgennanten Gebiete vorkommenden Gramineen, Cyperaceen und Juncaceen, mit . . . Werthe der einzelnen Arten fuer dio Landwirth- schaft, *etc.* Weimar, 1877. 8°.

Deutschland's Giftpflanzen. Hamburg, 1881. 8°.

HEINITSH (Charles A.).

Note on the culture of Saffron in Pennsylvania. [*Extract.*] *Pharm. Journ.* II. ix. (1867) 28–29.

HEINRICH (Fr.).

Ueber die Bestimmung reducirender Zucker neben Rohrzucker. (Sachsse, *Phytochemische Untersuchungen*, i. (1880) pp. 93– 100.)

HELDREICH (Theodor von).

Die Nutzpflanzen Griechenlands. Mit besonderer Beruecksich- tung der neugriechischen und pelasgischen Vulgarnamen. Athen, 1862. 8°.

HELDRING (L. J.).

De behandeling van de kaspischen wilg (Salix acutifolia) op hooge zandgronden. 2e vèrbet. druk. Arnhem, 1878. 8°.

—— 2de druk, met een voorwoord benevens eenige aanvullingen en verbeteringen. Aldaar, 1878. 8°.

HÉLOT (—.).

On the manufacture of Chinese green dye, called Lo-kao. *Pharm. Journ.* xvi. (1857) 517–520, 553–555.

HENDERSON (—.).

Cyrenaican drugs. [Extract from a consular report.] *Pharm. Journ.* III. iv. (1874) 598.

HENDERSON (*Dr.*).

Pharmaceutical results of the Calcutta Botanic Gardens. Accli- matization of Jalap and Ipecacuanha. [Extract from Dr. Henderson's report.] *Pharm. Journ.* III. iv. (1873) 221–222, 241–242, 261–263.

HENDRICH (J.).

Die wichtigsten landwirthschaftlichen Culturpflanzen, zum praktischen Gebrauche fuer Landwirthe, Taxatoren, Ackerbau-
schueler, *etc.* Prag, 1878. 4°.

Specieller Pflanzenbau. Kurze Anleitung zum Anbau der landwirthschaftlichen Kulturpflanzen. Prag, 1880. 8°.

HENKEL (J. B.).

Systematische characteristik der medicinisch wichtigen Pflanzenfamilien. Wuerzburg, 1856. 12°.

HENKEL (J. B.), & W. HOCHSTETTER.

Synopsis der Nadelhoelzer, deren charakteristische Merkmale
nebst Andeutungen ueber ihre Cultur und Ausdauer in
Deutschlands Klima. Stuttgart, 1864. 8°.

HENRIECK (G. A.).

Du tabac, son histoire, sa culture, sa fabrication, son commerce,
etc. Paris, 1866. 18°.

HENRY (Aimé).

Die Giftpflanzen Deutschlands zum Schulgebrauch und Selbstunterricht durch Abbildungen und Beschreibungen erlaeutert.
Bonn, 1836. 8°.

HENRY (J.).

Les principales plantes vénéneuses. Bruxelles, 1881. 8°.

HÉRAUD (A.).

Nouveau dictionnaire des plantes médicinales. Description, habitat
et culture, récolte, *etc.* Paris, 1875. 18°.

Nuevo diccionario de las plantas medicinales estudiados bajo cl
punto de vista botánico, medico y farmaceutico. Traducido
y adicionado con los nombres vulgares españoles de las plantas
medicinales que se encuantran en España, por J. G. Hidalgo.
Madrid, 1876. 4°.

HERDER (Ferdinand G. von).

Verzeichniss saemtlicher botanischen und landwirthschaftlichen
Gaerten, sowie der botanischen Museen, Herbarien und
verwandten Institute in allen fuenf Welttheilen, mit Angabe
ihres derzeitigen Vorstandpersonals, nach den einzelnen Staaten
in alphabetischer Reihenfolge zusammengestellt. St. Petersburg, 1870. 8°.

HÉRINCQ (F.).

Observations critiques sur l'origine des plantes domestiques.
[Extracts from Nos. 5, 6, and 7 (1869) of *L'Horticulteur
français.*] Paris, 1869. 8°.

HÉRINCQ (F.) *continued :—*

La verité sur le protendu silphion de la Cyrenaïque (Silphium cyrenaïcum de Laval). Ce qu'il est, ce qu'il n'est pas. 2e edition. Paris, 1876. 8°.

Silphion of the ancients and its alleged modern representative. [Abstract.] *Pharm. Journ.* III. vii. (1877) 750–753.

HERLANT (A.).

Études sur les principaux produits résineux de la famille des Conifères. Bruxelles, 1876. 8°.

HERMANN (Paul).

Der Pilzjaeger, odor die in Deutschland wachsenden essbaren, verdaechtigen oder nicht essbaren und schaedlichen Pilze. Ed. 2. Dresden, 1854. 8°.

HERRERA (Alfonso).

Notes on the Joyote of Mexico. [*Amer. Journ. Pharm.*] *Pharm. Journ.* III. vii. (1877) 854–855.

HERRMANN (Karl Robert).

Oekonomische Pflanzenkunde der landwirthschaftlichen Kultur-gewaechse. Colberg, 1846–47. 8°.

HERSCHEL (*Sir* John Frederick William).

Botanical and Pharmaceutical desiderata. [From the *Manual of Scientific Enquiry*, edited by Sir F. W. Herschel, Bart.] Africa (including Arabia and Abyssinia), *Pharm. Journ.* ix. (1849) 285–286. Asia (including Australia), *Pharm. Journ.* ix. (1850) 334–337. America, *Pharm. Journ.* ix. (1850) 585–586.

         Manual, ed. 2. 1851 ; Ed. 3. by R. Main, 1859 ; Ed. 4. 1871.

         See also HANBURY & OLIVER ; and HOOKER (W. J.).

HESS (Ch.).

Zusammenstellung der vorzueglichsten essbaren und giftige Schwaemme des Fichtelgebirges. Programm. Baireuth, 1843. 4°.

HESSE (O.).

Remarks on the Javanese Calisaya and on Conchinine. [Extract.] *Pharm. Journ.* III. v. (1874) 482–483.

Beitrag zur Kenntniss der Chinarinden. (Liebig's *Ann. der Chemie,* cc. (1880) 302–310.)

Zur Kenntniss der Pereirorinde. (Liebig's *Ann. der Chem.* ccii. (1880) 141–149.)

HEURCK (Henri van).

Du Boldo [Boldu chilanum, *Nees.*] Bruxelles, 1873. 8°.

Du Jaborandi. [Pilocarpus pennatifolius.] Bruxelles, 1875. 8°.

HEURCK (Henri van), & Victor GUIBERT.

Flore médicale belge. Louvain, 1864. 8°.

HEUZÉ (Gustave).

Plantes fourragères. Versailles, 1856. 8°. Ed. 3. 1861.

Les plantes industrielles. Les plantes oléagineuses (colza, navette, pavot-oeilette, cameline, ricin, arachide, sésame, etc.) Paris, 1859–60. 8°. Ed. 2. 1869. 18°.

Les plantes alimentaires. Ouvrage accompagné d'un atlas contenant 102 épis de cereales de grandeur naturelle dessinés par L. Rouyer, *etc.* Paris, 1873. 8°.

Les céréales, les produits farineux et leurs derivés à l'Exposition universelle internationale de 1878 à Paris. Paris, 1881. 8°.

HEWETT (Charles).

Chocolate and Cocoa. Cocoa; its growth and culture, manufacture, *etc.* London, 1862. 8°. [Ed. 2.] 1864.

HEWETT (J. F. Napier).

European settlements on the West Coast of Africa, with remarks on the supply of Cotton. London, 1862. 8°.

HIBBERD (Shirley).

On a new system of cultivating the Potato, with a view to augment production and prevent disease. *Journ. Soc. Arts,* xxii. (1874) 293–297.

The cultivation of hardy fruits, with a view to improvement of quality and insuring constant and abundant production. *Journ. Soc. Arts,* xxiv. (1876) 198–203.

HIGGINS (Bryan).

Observations and advices for the improvements of the manufacture of Muscovado Sugar and Rum. St. Jago de la Vega, 1797–1800. 2 pts. 8°.

HILDT (Johann Adolph).

Beschreibung in- und auslaendischer Holzarten. Weimar, 1798–99. 2 vols. 8°.

HILL (John), *styling himself* Sir John Hill.

The useful Family Herbal, *etc.* Ed. 2. London, 1755. 8°. [Ed. 3.] Bungay, 1812.

The British Herbal: an history of plants and trees, natives of Britain, cultivated for use or raised for beauty. London, 1756. fol.

Virtues of British herbs, with the history, description, and figures, *etc.* Ed. 4. London, 1771. 8°.

HIRSCHSOHN (Eduard).

Étude comparative du galbanum et de la pomme ammoniaque, suivie de quelques considerations sur l'opoponax et le saga-pénum, traduit de l'allemand par la docteur Jul. Morel. Gand. 1876. 8°.

HIRSCHSOHN (Eduard) *continued :—*

    Comparative examination of the more important commercial varie-
ties of Galbanum and Ammoniacum gums. [Extract.] I. Galbanum. *Pharm. Journ.* III. vii. (1876) 369–371, 389–391,
429–431, 531–533, (1877) 571–572. II. Ammoniacum.
612–614, 710–712. III. Comparison of Galbanum with Ammoniacum, 770–771. IV. Sagapenum and Opoponax, 771–
772.

HIRTH DU FRENES (F.).

    Commercial drugs of the Chinese province of Kuang-tung
(Canton). [Extract.] *Pharm. Journ.* III. viii. (1877) 88,
345.

HISTED (Edward).

    On artificial flake manna. *Pharm. Journ.* II. xi. (1870) 629–
630. *See also* HANBURY.

HOCHSTETTER (Christian Friederich).

    Die Giftgewaechse Deutschlands und der Schweiz. Esslingen,
1844. 8°

HODGES (John Frederick).

    The raw material of the linen trade : Flax. Belfast, 1865. 8°.

HOEHNEL (Franz von).

    Die Gerberrinden. . Ein monographischer Beitrag zur technischen
Rohstofflehre. Berlin, 1880. 8°.

HOEYELSEN (F. Chr.).

    Kort Anviisning til at avle Kummen. Kjoebenhavn, 1824.
8°.

HOESS (Franz).

    Monographie der Schwarzfoehre (Pinus austriaca) in botanischer
und forstlicher Beziehung. Wien, 1831. fol.

HOFFER (R.).

    Kautschuk und Guttapercha. Wien, 1880. 8°.

HOFFMANN (J.).

    Die Angaben schinesischer und japanischer Naturgeschichten
von dem Illicium religiosum (dem Mangsthao der Schinesen,
Sikiminoki der Japaner) und dem davon verschiedenen Stern-
anis des Handels. Leiden, 1837. 8°.

HOFFMANN (M.).

    Die offizinellen Gewaechse der preussischen Pharmacopoee fuer
Mediciner Pharmaceuten und Droguisten bildlich dargestellt
und beschrieben. Jena, 1863–64. 8°.

HOFMANN (F. W.).

    Die Cultur der Handelsgewaechse. Prag, 1835. 8°.

HOGENDORP (Dirk van).

Berigt van den tegenwoordigen toestand der Bataafsche bezit-
tingen in Oost-Indiën en den handel op dezelve. [Delft,
1799.] 8°.

HOGG (Robert).

The Vegetable Kingdom and its products, *etc.* London, 18[57–]
58. 8°.

HOHENBRUCK (Arthur von).

Der Holzexport Oesterreichs. Wien, 1869. 8°.

Beitraege zur Statistik des Flachs- und Hanf Production in
Oesterreich. Wien, 1874. 4°.

HOLL (Friedrich).

Die Verwechselungen und Aehnlichkeiten der wichtigsten
officinellen Pflanzen. Dresden, 1835. 4°.

Woerterbuch deutscher Pflanzennamen oder Verzeichniss
saemmtlicher in der Pharmacie, Oekonomie, Gaërtnerei,
Forstkultur und Technik vorkommenden Pflanzen. Erfurt,
1833. 8°.

Spurious Cascarilla bark. [From the *Pharm. Centr. Blatt.*]
Pharm. Journ. vii. (1847) 35.

On Cusco bark. *Pharm. Journ.* vii. (1847) 351.

HOLL (Friedrich), & Gustav HEYNHOLD.

Flora von Sachsen . . . mit besondrer Beruecksichtung ihrer
Verwendung. (Vol. i.) Dresden, 1842–43. 8°.

HOLLE (K. F.).

Proef-handleiding voor de kultuur en gewone inlandsche bereiding
van koffij. Batavia, 1863. 8°.

HOLM (John).

On Cocoa and its Manufacture. *Journ. Soc. Arts*, xxii. (1874)
356. Also in *Pharm. Journ.* III. iv. (1874) 804–806, 843–846,
885–886.

Cocoa and its manufacture, *etc.* London, 1874. 8°.

HOLMES (Edward Morell).

Materia medica of the United States Pharmacopoeia. *Pharm.
Journ.* III. iii. (1873) 1009–1011, 1029–1031; iv. 123–127,
162–163, 201–202.

Vegetable poisons and their antidotes. *Pharm. Journ.* III. iv.
(1874) 769–770.

Materia medica notes. Koegoed. *Pharm. Journ.* III. iv. (1874)
810–811. *See also* p. 821.

The botanical source of Jaborandi. [Pilocarpus sp.] *Pharm.
Journ.* III. v. (1875) 581–583.

HOLMES (Edward Morell) *continued :—*

Note on a spurious Senna. *Pharm. Journ.* III. v. (1875) 623–624. *See also* p. 634.

Further note on the botanical source of Jaborandi. *Pharm. Journ.* III. v. (1875) 641.

A second kind of Jaborandi. *Pharm. Journ.* III. v. (1875) 781–782.

On the identity of Goa powder and Araroba. *Pharm. Journ.* III. v. (1875) 801–802. *See also* pp. 816–818.

Notes on Brazilian drugs. *Pharm. Journ.* III. v. (1875) 905–906, 985–986.

Gelsemium sempervirens. *Pharm. Journ.* III. vi. (1875) 481–482, 521–522, (1876) 561–564, 601–602.

The botanical source of Damiana. *Pharm. Journ.* III. vi. (1876) 581.

Notes on some American medicinal plants. *Pharm. Journ.* III. vi. (1876) 781–782.

On adulteration of Aconite root. *Pharm. Journ.* III. vii. (1877) 749–750.

The cultivation of medicinal plants at Banbury. *Pharm. Journ.* III. vii. (1877) 1017–1019.

Note on Rheum officinale (Baill.). *Pharm. Journ.* III. viii. (1877) 181–182.

The cultivation of medicinal plants at Hitchin. *Pharm. Journ.* III. viii. (1877) 301–303.

Notes on casual drugs. *Pharm. Journ.* III. viii. (1877) 363–364. *See also* p. 378.

Notes on the medicinal plants of Liberia. *Pharm. Journ.* III. viii. (1878) 563–564.

Duboisia myoporoides, *R. Br. Pharm. Journ.* III. viii. (1878) 705–706. *See also* p. 720.

Note on Grindelia robusta. *Pharm. Journ.* III. viii. (1878) 787. *See also* pp. 797–798.

On adulteration of Senega. *Pharm. Journ.* III. ix. (1878) 410–411. *See also* pp. 419–420.

Guaycuru root. [Statice brasiliensis?] *Pharm. Journ.* III. ix. (1878) 466. *See also* p. 481.

Notes on some Japanese drugs. *Pharm. Journ·* III. x. (1879) 3–5, 21–23, 101–103, 201–203, 261–262.

Myrtus Chekan. *Pharm. Journ.* III. ix. (1879) 653–654.

Note on Shea butter. [Butyrospermum Parkii.] *Pharm. Journ.* III. ix. (1879) 818–819.

HOLMES (Edward Morell) *continued :*—
Further notes on Liberian drugs. *Pharm. Journ.* III. ix. (1879) 853.

Note on Calabar beans. *Pharm. Journ.* III. ix. (1879) 913–914.

Japanese Belladonna. *Pharm. Journ.* III. x (1880) 789.

The botanical source of Tonga. *Pharm. Journ.* III. x. (1880) 889.

The use of Saffron in the Pharmacopoeia. *Pharm. Journ.* III. xi. (1880) 450–451. *See also* pp. 461–463.

Star-anise. *Pharm. Journ.* III. xi. (1880) 489–491.

Jafferabad Aloes. *Pharm. Journ.* III. xi. (1881) 733. *See also* pp. 746–748.

HOLMES (*Sir* William Henry).
Free Cotton; how and where to grow it. With a map of British Guiana. London, 1862. 8°.

HOLMGREN (August Emil).
Anvisning att igenkaenna Sweriges vigtigara loeftrad och loef-buskar under deras blad- och bloemloesa tillstånd jemto en korte framstaellning af traedslagens naturliga familjer. Stockholm, 1861. 8°.

HOLSTEIN (Fr. Adolf von).
Anviisning for Gaardmaendene i Grevskabet Holsteinborg til at opelske Skov paa deres Lodder. Soroee, 1811. 8°.

HOLZNER (Georg).
Die Beobachtungen ueber die Schuette der Kiefer oder Foehre und die Winterfaerbung immergruener Gewaechse. Fuer Forstmaenner und Botaniker zusammengestellt nebst Bemer-kungen. Freising, 1878. 8°.

HÓMAN (Bálint).
A szíkes talaj műveleséröl és fatenyésyetéröl [Culture, and tree vegetation of soda-containing soils.] *Erdészeti lapok*, 1880, No. 12, 925–928.

HONIGBERGER (Johann Martin).
Fruechte aus dem Morgenlande, oder Reise-Erlebnisse, nebst Naturhistorisch-medicinischen Erfahrungen, einigen hundert erprobten Arzneimitteln, und einer neuen Heilart, dem Medial-Systeme. Wien, 1852. 8°.

Thirty-five years in the East. Adventures, Discoveries, Expe-riments, and Historical Sketches, relating to the Punjab and Cashmere; in connexion with Medicine, Botany, Pharmacy, etc., together with an original Materia medica; and a medical vocabulary in four European and five Eastern languages. London, 1852. 8°.

HOOD (John W.).

Notes on the cultivation of the Opium Poppy in Australia. *Pharm. Journ.* III. i. (1870) 272–274.

HOOKER (*Dr.*, afterwards *Sir* Joseph Dalton).

Description of a new species of Amomum [A. Danielli] from Tropical West Africa. [From Hooker's *Journ. Bot.*] With additional note by J. Pereira. *Pharm. Journ.* xii. (1852) 72–73.

HOOKER (*Sir* William Jackson).

Botanical characters of a new plant (Isonandra pulla) yielding the Gutta Percha of Commerce. [From the *Lond. Journ. Bot.*] *Pharm. Journ.* vii. (1847) 179–181.

On Patchouli. [Pogostemon Patchouli, *Lepell.* = P. intermedius, *Benth.*, an earlier name, but accidentally omitted from DC. Prod.] *Pharm. Journ.* ix. (282).

Some account of the Vegetable Ivory Palm. (Phytelephas macrocarpa.) [From Hooker's *Journ. Bot.*] *Pharm. Journ.* ix. (1850) 369–375.

On Piaçaba, and Coquilla nuts. [Attalea funifera. From Hooker's *Journ. Bot.*] *Pharm. Journ.* ix. (1850) 431–432.

Jute: Corchorus capsularis, L. [From Hooker's *Journ. Bot.*] *Pharm. Journ.* ix. (1850) 545.

Chinese "Rice-Paper," or "Bok-Shung." [From Hooker's *Journ. Bot.*] *Pharm. Journ.* ix. (1850) 545–546.

On Boehmeria nivea, or Chinese Grass-plant, and on Boehmeria Puya. *Pharm. Journ.* xi. (1851) 276–278.

Description and figure of the Cedron of the Magdalena River. (Simaba Cedron, *Planchon.*) [From Hooker's *Journ. Bot.*] *Pharm. Journ.* x. (1851) 344–348.

Cedron. [From Hooker's *Journ. Bot.*] *Pharm. Journ.* x. (1851) 472.

Amomum Granum-paradisi. Grains of Paradise Amomum; or Mellegetta pepper. [From the *Botanical Magazine.*] *Pharm. Journ.* xii. (1852) 192–194.

Coscinium fenestratum (False Calumba-root). [From Hooker's *Journ. Bot.*] *Pharm. Journ.* xii. (1852) 185–188.

On the Camphor-tree of Borneo and Sumatra, Dryobalanops Camphora, *Colebr.* [From Hooker's *Journ. Bot.*] *Pharm. Journ.* xii. (1852) 300–302.

Museum of Economic Botany, or a popular guide to the Museum of the Royal Gardens of Kew. London, 1855. 8°.

For subsequent editions, see OLIVER (D.).

HOOKER (*Sir* William Jackson) *continued :—*
  Report on vegetable products obtained without cultivation. (Paris Universal Exhibition.) London, 1857. 8°.
  Commercial products of the Asphodel. [From *The Technologist.*] *Pharm. Journ.* II. iii. (1861) 24.
HOOKER (*Sir* William Jackson), & Daniel HANBURY.
  Botanical and Pharmacological Inquiries and desiderata. [From the *Admiralty Manual of Scientific Enquiry. See also* HERSCHEL (J. F. W.).] *Pharm. Journ.* II. i. (1859) 217–220.
HOPE (H. W.).
  Essai sur l'exploitation du chêne-liège en algérie. Paris, 1876. 8°.
HORNE (J.).
  Fiji. Remarks on the agricultural prospects of Fiji. Levuka, 1878. 8°
HORNER (W. H.).
  Cotton in Missouri. *Report, Commissioner of Patents,* Agriculture, 1861, (Washington, 1862) 221–223.
HOUGH (Franklin B.).
  Report upon Forestry, prepared under the direction of the Commissioners of Agriculture. Washington, 1878. 8°. *
HOULTON (Joseph).
  On annual and biennial Hyoscyamus. *Pharm. Journ.* ii. (1842) 68.
  Hyoscyamus and Digitalis. [Difference in leaves of plants of various ages.] *Pharm. Journ.* i. (1842) 406–407.
  Of the medicinal properties of Leontodon Taxacum. From the "Medical and Surgical Journal," vol. i. 1828. *Pharm. Journ.* i. (1842) 421–424.
  Hyoscyamus niger. [A paper read to the Royal Medico-Botanical Society.] *Pharm. Journ.* iii. (1844) 578–581.
  On Digitalis purpurea. The substance of a paper read before the Medico-Botanical Society. *Pharm. Journ.* iv. (1844) 126–128.
HOUSSAYE (J. G.).
  Monographie du thé, description botanique, torréfaction, composition chimique, propriétés hygiéniques de cette feuille. Paris, 1843. 8°. [A list of 58 authors who have written on Tea, from 1590 to 1843, is given on pp. 155–157.]
HOUSTON (Joseph).
  Collection and preservation of Hyoscyamus. From the *Lancet,* May 8th, 1841. *Pharm. Journ.* i. (1842) 427–428.
  Melilotus caerulea. *Pharm. Journ.* ii. (1842) 128.
  Colchicum autumnale. *Pharm. Journ.* ii. (1842) 129[–130].

HOUTTUYN (Martin).

Houtkunde ; verzameling van in- en uitlandsche houten, en derzelver benamingen in het hollandsche, hoogduitsch, engelsch, fransch, en latijn, *etc.* Amstelaedami, 1773. 4°. Also in 1791 and 1795.

> The title is given in Dutch, German, French, English and Latin. A German ed. also, Nuernberg, 1773-98.

HOWARD (David).

Some recent importations of Cinchona barks. *Pharm. Journ.* III. v. (1875) 1025.

Note on the distribution of the alkaloids in Cinchona trees. *Pharm. Journ.* III. viii. (1877) 1-2.

Notes on Cinchona bark. *Pharm. Journ.* III. x. (1879) 181.

HOWARD (John Eliot).

Examination of Pavon's collection of Peruvian barks contained in the British Museum. *Pharm. Journ.* xi. (1852) 489-498, 557-564, xii. 11-16, 58-62, 125-129, 173-180, 230-235, xii. (1853) 339-342.

Observations on the specimens of Peruvian bark presented to the Museum of the Pharmaceutical Society, May 17th, 1854. *Pharm. Journ.* xiii. (1854) 671-672.

Observations on some additional specimens of Peruvian barks presented to the Museum of the Pharmaceutical Society. *Pharm. Journ.* xiv. (1854) 61-63.

On the bark of Gomphosia chlorantha, *Wedd.,* occurring mixed with quilled Calisaya bark. *Pharm. Journ.* xiv. (1855) 318.

On Copalche bark. *Pharm. Journ.* xiv. (1855) 319.

On the tree-producing red Cinchona bark. *Pharm. Journ.* xvi. (1856) 207-212.

Illustrations of the Nueva quinologie of Pavon, *etc.* London, 18[59-]62. fol.

Report on the bark and leaves of Chinchona succirubra grown in India. To the Under-Secretary of State for India. *Pharm. Journ.* II. v. (1863) 74-75.

New features in the supply of Peruvian bark. *Pharm. Journ.* II. v. (1863) 248-249.

On the root-bark of the Chinchonae. *Pharm. Journ.* II. vi. (1864) 19-21.

On the red variety of Pitayo bark. *Pharm. Journ.* II. vi. (1864) 48-50.

Note on the root-bark of Chinchona Calisaya. *Pharm. Journ.* II. v. (1864) 342-345.

HOWARD (John Eliot) *continued :*—

Microscopical researches on the alkaloids, as existing in Chinchona bark. *Pharm. Journ.* II. vi. (1865) 584–588. [With 4 plates.]

Observations on the present state of our knowledge of the genus Cinchona. (Abstract of a paper read at a meeting of the International Botanical Congress, and prepared by the author, by desire, for the Pharmaceutical Journal.) *Pharm. Journ.* II. viii. (1866) 11–16.

Report of an analysis of the sixth remittance of bark from India. *Pharm. Journ.* II. ix. (1867) 243–244.

Report of an analysis of the seventh remittance of bark from India. *Pharm. Journ.* II. ix. (1867) 245–246.

Report of an analysis of the eighth remittance of bark from India. *Pharm. Journ.* II. x. (1868) 317–320.

The Quinology of the East Indian plantations. London, 1869–76. fol.

On the cultivation of Cinchona plants under glass in England. *Pharm. Journ.* II. xi. (1870) 388–392.

Cinchona cultivation in Java. *Pharm. Journ.* III. i. (1870) 441– 442. *See also* p. 466.

Cinchona-trees grown in India. *Pharm. Journ.* III. ii. (1871) 361–363. *See also* pp. 374–375.

Correspondence relative to Cinchona cultivation in India. *Pharm. Journ.* III. ii. (1872) 724–727.

Report on Cinchona bark grown in Jamaica. *Pharm. Journ.* III. iii. (1872) 83.

Examination of the leaves of Cinchona succirubra. (Especially in reference to the production of alkaloid.) *Pharm. Journ.* III. iii. (1873) 541–542.

Cinchona cultivation in India. [Abstract from *The Gardeners' Chronicle.*] *Pharm. Journ.* III. iii. (1873) 647–648.

Cinchonas. [Abstract, *Linn. Soc. Journ.*] *Pharm. Journ.* III. iii. (1873) 881.

The Cinchona plantations in Java. *Pharm. Journ.* III. iv. (1873) 21–25, 41.

Presumed hybrid between Cinchona Calisaya and C. succirubra. *Pharm. Journ.* III. v. (1874) 1–2.

On coppicing Cinchonas. *Pharm. Journ.* III. iv. (1874) 797–798.

Indian barks contributed to the Museum. *Pharm. Journ.* III. v. (1875) 1005–1006.

The supply of Cinchona bark as connected with the present price of Quinine. *Pharm. Journ.* III. viii. (1877) 207–210.

HOWARD (John Eliot) *continued :—*

The fast-growing variety of Cinchona called pubescens. *Pharm. Journ.* III. viii. (1878) 825.

Origin of the Calisaya Ledgeriana of Commerce. *Pharm. Journ.* III. x. (1880) 730.

The cultivation of Calisaya. *Pharm. Journ.* III. xi. (1880) 244–246.

HUBER (Candidus).

Vollstaendige Naturgeschichte aller in Deutschland einheimischen und nationalisirten Bau- und Baumhoelzer. Muenchen, 1808. 2 vols. 4°.

HUEBNER (J. G.).

Pflanzen-Atlas. Koepenick, 1861. 4°. Ed. 2. Berlin, 1863. Ed. 3. 1869.

—— Planten-atlas. Voor inrigtingen van middelbaar onderwijs, aanstaande geneeskundigen en apothekers. Naar de 2. hoogduitsche uitgave door Corstiaan de Jong. Arnhem, 1864. 8°.

HUGHES (H. G.).

Tutu [Coriaria ruscifolia] as a dye wood. *Pharm. Journ.* III. ii. (1871) 303.

HUGHES (William).

The American physician, or a treatise of the roots, plants, trees, shrubs, fruits, herbs, . . . growing in the English plantations in America, *etc.* London, 1672. 12°.

HULL (Edmund C. P.).

Coffee, its physiology, history, and cultivation, adapted as a work of reference for Ceylon and the Neilgherries. Madras, 1865. 8°.

Coffee planting in Southern India and Ceylon. Being a second edition, enlarged, of " Coffee, its physiology," *etc.* London, 1877. 8°.

HULL (Hugh Munro).

On Tasmania and its wealth in timber. *Proc. Royal Colonial Inst.* iv. (1873) 169–173.

The economic value of the forests of Tasmania. *Proc. Royal Colonial Inst.* v. (1874) 160–165.

HULLE (H. J. van).

Guide arboricole aux cours publics de taille et aux écoles normales et primères. Ed. 2. Gand, 1875. 12°.

HUNDESHAGEN (Johann Christian).

Lehrbuch der forst- und landwirthschaftlichen Naturkunde, *etc.* Tuebingen, 1829. 8°.

Encyclopaedie der Forstwissenschaften. Tuebingen, 1842–43. 4 vols. 8°.

Hunter (Alexander).

On the fixed vegetable oils of Southern India. *Pharm. Journ.* xii. (1853) 598–599.

Hunter (Charles).

A report upon some of the colonial medicinal contributions to the International Exhibition, A.D. 1862. London, 1863. 8°.

Hurlbert (J. Beaufort).

The climates, productions, and resources of Canada. Montreal, 1872. 8°.

Husemann (August & Theodor).

Die Pflanzenstoffe in chemischer, physiologischer, pharmakologischer und toxicologischer Hinsicht, *etc.* Berlin, [1864–71]. 8°.

Husemann (Theodor).

The value of mushrooms as food. [Extract.] *Pharm. Journ.* III. viii. (1877) 85.

Java Rhubarb. [Extract.] *Pharm. Journ.* III. viii. (1877) 328–329.

The distribution of Cardiac poisons in the vegetable kingdom. [Abstract.] *Pharm. Journ.* III. vii. (1877) 795–796, 834–835.

Poisonous property of a "Star anise." [From the *Pharm. Zeitung.*] *Pharm. Journ.* III. xi. (1880) 453–454.

Huss (Magnus).

Om kaffe; dess bruk och missbruk, *etc.* Stockholm, 1865. 8°.

Huth (Johann Christopher).

Dissertatio medica de potu chocolatae, *etc.* [M. Mappus, *Praes.*] Argentorati, [1695]. 4°.

Hutton (F. W.).

A Lecture on the manufacture of New Zealand Flax. Auckland, 1870. 8°.

Icery (E.). *Edmond*

De quelques recherches sur le jus de la canne à sucre et sur les modifications qu'il subit pendant le travail d'extraction a l'ile Maurice. Maurice, 1865. 8°. *Paris*

Imbert-Courbeyre (A.).

Leçons sur le tabac, faites au palais des facultés de Clermont-Ferrand. Clermont-Ferrand, 1866. 12°.

Ince (Joseph).

Acclimatization. Part II. The Garden of Acclimatization. (Paris). *Pharm. Journ.* II. v. (1863) 28–32, 60–66.

INGRAM (W.).

Sorghum; a report of an experiment in its cultivation at Belvoir. London, 1881. 12°.

INZENGA (Giuseppe).

Manuale pratico delle coltivazione del sommacco in Sicilia. Palermo, 1875. 8°.

IRVINE (Robert Hamilton).

A short account of the Materia medica of Patna. Calcutta, 1848. 8°.

ISHIKAWA (J.).

Materials containing Tannin used in Japan. *Chemical News*, vol. xlii. (1880) pp. 274–277.

IVERSEN (Jac.).

Om Rapsaedens Dyrkning i det Holsteenske, i Saerdeleshed i Hertugdoemmet Slesvig. . . . Oversat af det Tydske. Kjoebenhavn, 1803. 8°.

Der Rapsaat-Bau in Hollsteinischen, besonders im Herzogthum Schleswig, *etc.* Bremen, 1806. 8°.

Ist der Rapsaatbau auf der Geest wirklich so bedenklich dasz wir ihn aufgegeben muessen? und wirkt der Mergel nur auf einen Saaten-Umlauf? Beantwortet. Augustenburg, 1821. 8°.

JACKSON (John Reader).

Cork and its uses. [From *The Technologist*.] *Pharm. Journ.* II. vi. (1865) 652–655.

Pepper. [From *The Technologist*.] *Pharm. Journ.* II. vii. (1865) 288–291.

Orchid Tea. [From *The Gardeners' Chronicle*.] *Pharm. Journ.* II. viii. (1866) 28–29.

Economic value of the common Brake. [Extract.] *Pharm. Journ.* II. viii. (1866) 354–355.

Ginseng. [From *The Gardeners' Chron.*] *Pharm. Journ.* III. i. (1870) 208–209.

Ginseng. *Pharm. Journ.* III. i. (1871) 665.

The cultivation of Opium in China. *Pharm. Journ.* III. i. (1871) 782–783.

The Tamarind. *Pharm. Journ.* III. i. (1871) 863.

The Ochro and the Musk Mallow. *Pharm. Journ.* III. i. (1871) 965–966.

Notes on some Eastern varnish trees. *Pharm. Journ.* III. ii. (1871) 61.

The uses of the genus Cyperus. *Pharm. Journ.* III. ii. (1871) 502–503.

JACKSON (John Reader) *continued :—*

The cultivation and use of the Dandelion in India. *Pharm. Journ.* III. ii. (1871) 523–524.

The Algerian Callitris. *Pharm. Journ.* III. ii. (1872) 623.

Notes on the properties of the Geranieae. *Pharm. Journ.* III. ii. (1872) 744–745.

Poisonous properties of Jatropha urens. *Pharm. Journ.* III. ii. (1872) 863–864.

The so-called African saffron. *Pharm. Journ.* III. ii. (1872) 904.

The economic and medicinal value of the genus Rhus. *Pharm. Journ.* III. ii. (1872) 985.

The medicinal properties of the cow trees of South America. *Pharm. Journ.* III. iii. (1872) 321–322.

Blue mountain tea. [Solidago odora, *Ait.*] *Pharm. Journ.* III. iii. (1873) 603.

The medicinal plants of New Zealand. *Pharm. Journ.* III. iii. (1873) 662–663.

Churrus. *Pharm. Journ.* III. iii. (1873) 764.

Notes on the medicinal plants of the Rutaceae. *Pharm. Journ.* III. iii. (1873) 951–953.

African tea plants. *Pharm. Journ.* III. iv. (1873) 421.

Note on Liatris odoratissima. *Pharm. Journ.* III. iv. (1873) 322.

Notes on the Areca palm. Areca Catechu, *L. Pharm. Journ.* III. iv. (1874) 689.

Notes on the medicinal plants of the Scrophulariaceae. *Pharm. Journ.* III. iv. (1874) 1033–1034.

The Uses of Agave americana. *Pharm. Journ.* III. v. (1874) 461–462.

Ginseng. (Panax Schinseng of Nees.) [From *The Gardeners' Chronicle.*] *Pharm. Journ.* III. vi. (1875) 86–88.

Notes on some medicinal plants of the Compositae. *Pharm. Journ.* III. vi. (1875) 462–464.

Vanilla. *Pharm. Journ.* III. v. (1875) 885–886.

Zebra wood. [From *The Gardeners' Chronicle.*] *Pharm. Journ.* III. v. (1875) 1009.

Princewood bark, a febrifuge from the Bahamas. *Pharm. Journ.* III. vi. (1876) 681.

Another note on Rhubarb. *Pharm. Journ.* III. vi.·(1876) 966.

Notes on the drugs collected by the Prince of Wales in India. *Pharm. Journ.* III. vii. (1876) 129–130.

Jackson (John Reader) *continued :—*

Fenugreek. [From *The Gardeners' Chronicle.*] *Pharm. Journ.*
III. vii. (1876) 157.

"Chicle" gum and Monesia bark. *Pharm. Journ.* III. vii.
(1876) 409.

Notes on some of the pharmaceutical products exhibited in the
Philadelphia exhibition of 1876. *Pharm. Journ.* III. vii.
(1877) 997–998, 1037–1039.

Notes on the uses of a commercial cane termed "whangee,"
species of Phyllostachys. (*Linn. Soc. Journ. Bot.* xvi.)

Sanguinaire or the Arabe. [Algerian tea.] *Pharm. Journ.* III.
viii. (1878) 521.

The uses of some of the Indian species of Bassia. *Pharm. Journ.*
III. viii. (1878) 646–648.

Boa-tam-paijang (Sterculia scaphigera, *Wall.*). *Pharm. Journ.*
III. viii. (1878) 747.

Indian plants adapted for commercial purposes. *Journ. Soc. Arts,*
xxvii. (1879) 333–342.

Jacobson (J. J. L. J.).

Handboek voor de kultuur en fabrikatie van thee. Batavia,
1843. 3 vols. 8°.

Jacques (Antoine ?).

Monographie de la famille des Conifères. Paris, 1837. 8°.
See Pritzel, Thes. ed. 2. p. 423.

Jaeger (B.).

Lectures sur l'histoire naturelle d'Haiti, appliquée à l'économie
rurale et domestique. Tom. i. contenant la botanique. Port-
au-Prince, 1830. 4°.

Jaeger (H.).

Die ·Nutzholzpflanzungen und ihre Verwendung, mit besonderer
Ruecksicht auf fremde Holzarten und Weidenzucht . . .
Fuer Gutsbesitzer, Forstleute, Gemeinde- und Eisenbahnver-
waltungen und Gaertner. Hannover, 1877. 8°.

Jager (F.).

Pyrethrum roseum, a remedy for insect-bites. [Extract.]
Pharm. Journ. II. x. (1868) 47.

Jago (—.).

Olive cultivation in Syria. [Extract from Consular report.]
Pharm. Journ. III. iii. (1873) 607.

Jahn (C. L.). ·

Die Holzgewaechse des Friedrichshains bei Berlin. Ein Verzeichniss derselben, *etc.* Berlin, 1864. 8°.

JAMES (A. G. F. Eliot).
Indian Industries. London, 1880. 8°.

JAMIE (—.).
Michelia Champaca. [Extract from a letter.] *Pharm. Journ.* III.
iii. (1872) 572–573.

JANSSEN (J.).
Fraxinus cultivation and Manna production. [Extract.] *Pharm.*
Journ. III. x. (1879) 407.

JAUBERT (Hippolyte François).
La botanique à l'exposition universelle de 1855. Paris, 1855.
8°.

JENNINGS (James).
A practical treatise on the history, medical properties, and
cultivation of Tobacco. London, 1830. 12°.

JESSEN (Carl Friedrich Wilhelm).
Deutschlands Graeser und Getreidearten zu leichter Erkenntniss
nach dem Wuchse den Blaettern, Bluethen und Fruechten
zusammengestellt und fuer die Land- und Forstwissenschaft
nach Vorkommen und Nutzen ausfuehrlich beschrieben.
Leipzig, 1863. 8°.

JOBST (J.), & O. HESSE.
Dita bark. [Alstonia scholaris, *R. Br.* Abstract.] *Pharm.*
Journ. III. vi. (1875) 142–144.

JOERLIN (Engelbert).
Specimen botanico-oeconomicum de usu quarundam plantarum
indigenarum prae exoticis. Londini Gothorum, 1769. 4°.

JOHANSSON (J.).
Om nora skog. Aeldra och nyare anteckningar. Stockholm,
1875. 8°.

JOHNSON (Charles Pierpoint).
The useful plants of Great Britain; a treatise upon the principal
native vegetables capable of application as food, medicine, or
in the arts and manufactures. Illustrated by John E. Sowerby.
London, 1862. 8°.

JOHNSON (Samuel W.).
Landtbruket och naturvetenskaperna. I. De odlade vaexternas
beståndsdelar, utveckling och lifsvilkor. II. De odlade vaexternas foeda samt luftens och jordens forhållenden till vaext-
naeringen. (Transl. by E. Bergstrand.) Stockholm, 1874. 8°.

JOHNSTON (James Finlay Weir).
The Chemistry of Common Life. Edinburgh and London, 1855.
2 vols. 8°. Ed. [2.] 1859. Ed. [3.?] 1879.

JOHNSTON (Robert).

Esparto : a series of practical remarks on the nature, cultivation, past history, and future prospects of the plant; including a demonstration of the importance to the paper making trade of prompt and vigorous measures for its preservation. *Journ. Soc. Arts*, xx. (1871) 96–104.

JOLY (Nicolas).

Observations générales sur les plantes qui peuvent fournir des couleurs bleues à la teinture, suivies de recherches anatomiques, physiologiques et chimiques sur le Polygonum tinctorium et spécialement sur le Chrozophora tinctoria (Croton tinctorium, L.) Montpellier, 1839. 4°.

Études sur les plantes indigofères en général et particulièrement sur le Polygonum tinctorium. Montpellier, 1839. 8°.

JUCH (Karl Wilhelm).

Die Giftpflanzen. Zur Belehrung fuer Jedermann. Augsburg, 1817. fol.

JULIEN (Stanislas).

Résumé des principaux traités chinois sur la culture des muriers et l'education des vers à soie. Paris, 1837. 8°.

JUS (H.).

Les plantes textiles algériennes à l'exposition universelle de 1878. Histoire d'une botte d'Alfa. Batna, 1878. 8°.

JUSSIEU (Adrien de).

De Euphorbiacearum generibus medicisque earundem viribus tentamen. Parisiis, 1824. 4°.

K.

On a newly discovered plant yielding the Gum Ammoniacum. (Diserneston gummiferum, *Jaub. & Spach*.) [From the *Pharm. Centr. Blatt*, April, 12, 1843.] *Pharm. Journ.* ii. (1843) 773.

KALBRUNER (Hermann).

The insecticidal properties of some species of Pyrethrum. [Extract.] *Pharm. Journ.* III. v. (1874) 503–504.

KAMPHAUSEN (N. W.).

Leitfaden zur Zucht des Maulbeerbaumes und der Seidenraupen. Coblenz, 1864. 8°.

KARSTEN (H.).

Deutsche Flora. Pharmaceutisch-medicinische Botanik. Ein Grundriss der systematischen Botanik zum Selbststudium fuer Aerzte, Apotheker und Botaniker. Berlin, 1880. 8°.—>

KATZER (Joseph).

Systematische Uebersicht der officinellen Pflanzen, welche in der Oestreich'schen Pharmacopoe enthalten sind. Wien, 1840. 8°.

KAYSER (G. A.).

On the two varieties of Jalap in commerce. [From the *Ann. der Chemie und Pharm.*] *Pharm. Journ.* iv. (1844) 284, 285 ; (1845) 327.

On the composition of the Jalap resin, obtained from the genuine Jalap tuber. (The Ipomœa Schiedeana of Zuccharini.) [From the *Ann. der Chemie und Pharm.*] *Pharm. Journ.* iv. (1845) 428–430.

KEENE (P. T.).

Botanical description of Cundurango. [Extract.] *Pharm. Journ.* III. ii. (1871) 405.

KELCH (Wilhelm Gottlieb).

Flora medica borussica, sistens plantas officinales sponte vigentes. Regiomonti, 1805. 8°.

KEMP (David S.).

On Goa powder. *Pharm. Journ.* II. v. (1864) 345–347.

Chrysarobine, or Goa powder. *Pharm. Journ.* II. v. (1875) 729.

KEMPTHORNE (G. B.).

Description of the Frankincense-tree, as found near Cape Gardafoi, on the Somauli Coast. [From Sir W. C. Harris's *Highlands of Ethiopia.*] *Pharm. Journ.* iv. (1844) 37, 38.

KENNEDY (George W.).

Note on Opium culture. *Pharm. Journ.* III. i. (1871) 762.

KERCHOVE DE DENTERGHEM (Oswald de).

Les palmiers, histoire iconographique. Géographie, paléontologie, botanique, description, culture, emploi . . . avec index général des noms et synonyms des espèces connues, *etc.* Paris, 1878. 8°.

KERNER (Johann Simon).

Handlungsprodukte aus dem Pflanzenreich. Stuttgart, 1781–86. fol.

Giftige und essbare Schwaemme, welche sowohl im Herzogthum Wirtemberg als auch ins uebrigen Teutschland wild wachsen. Stuttgart, 1786. 8°.

Abbildung aller oekonomischen Pflanzen. Figures des plantes économiques. Stuttgart, 1786–96. 8 vols. 4°.

KERR (Hem Chunder).

Report on the cultivation of and trade in Jute in Bengal, and on Indian fibres available for the manufacture of paper, with a map of the jute-growing districts of Bengal. Ed. 3. Calcutta, 1877. fol.

KERR (Thomas).

A practical treatise on the cultivation of the sugar cane, and the manufacture of sugar. London, 1851. 8°.

KEYWORTH (George Alexander).

Koegoed. *Pharm. Journ.* III. iii. (1872) 205.

KIENITZ (M.).

Ueber Formen und Abarten heimischer Waldbaeume. Berlin, 1879. 8°.

Schuessel zum Bestimmen der wichtigsten Deutschland cultivirten Hoelzer nach mit unbewaffnetem Ange erkennbaren Merk-malen. Muenden, 1880. 8°.

KING (George).

Cinchona plantations in British Sikkim. Tenth annual Report, *etc. Pharm. Journ.* III. iii. (1872) 302-304, 324-325.

Fourteenth annual report of the Government Cinchona plantation in British Sikhim. [Extract.] *Pharm. Journ.* III. vii. (1876) 512-513, 534-537.

A manual of Cinchona cultivation in India. Calcutta, 1876. fol. Ed. 2. 1880.

KINGZETT (Charles Thomas).

Nature's Hygiene : a series of Essays on popular scientific subjects, with special reference to the Chemistry and Hygiene of the Eucalyptus and the Pine. London, 1880. 8°.

KIRK (John).

On the Copal of Zanzibar. [*Journ. Linnean Soc.*] *Pharm. Journ.* II. x. (1869) 654-655.

KIRWAN (Charles de).

La France forestière depuis les temps les plus reculés jusqu'à nos jours. Paris, 1869. 8°.

Flore forestière illustrée ; arbres et arbustes du centre de l'Europe ; description générale, organographie, culture, habitat, produits principaux et accessoires. Paris, 1872. fol.

Notice sur l'industrie des écorces à tan. Paris, 1878. 4°.

KLEEMANN (G.).

Der praktische Zuckerruebenbau. Leipzig, 1881. 8°.

Koch (Karl Heinrich Emil).

Hortus dendrologicus. Verzeichniss der Baeume, Straeucher und Halbstraeucher, die in Europa, Nord- und Mittelasien, im Himalaya und in Nordamerika wild wachsen und moeglicher Weise in Mitteleuropa im Freien ausdauern; nach dem natuerlichen Systeme und mit Angabe aller Synonymie, sowie des Vaterlandes, aufgezaehlt und mit einem alphabetischen Register versehen. Berlin, 1853. 8°.

Dendrologie. Baeume, Straeucher und Halbstraeucher welche in Mittel- und Nord-Europa im Freien kultivirt werden kritisch beleuchtet. Erlangen, 1869–73. 8°.

Koch (Ludwig).

Die Klee- und Flachsseide (Cuscuta Epithymum und C. Epilinum). Untersuchungen ueber deren Entwickelung, Verbreitung und Vertilgung. Heidelberg, 1880. 8°.

Koenig (C. E. von).

Die Seradella, der Klee des Sandes. Wittenberg, 1864. 8°. Ed. 2. 1865. Ed. 3. 1873.

Kohlhaas (Johann Jakob).

Giftpflanzen auf Stein abgedruckt mit Beschreibungen. Regensburg, 1805. 4°.

Kohn (Ferdinand).

On the different methods of extracting sugar from beet-root and cane. *Journ. Soc. Arts,* xix. (1871) 338–345.

Kosteletzky (Vincenz Franz).

Allgemeine medizinisch pharmaceutische Flora, enthaltend die systematische Aufzaehlung und Beschreibung saemmtlicher bis jetzt bekannt gewordner Gewaechse aller Welttheile. Prag, 1831–36. 6 vols. 8°.

Kotschy (Theodor).

Die Eichen Europa's und des Orients. Gesammelt, zum Theil neu entdeckt, und mit Hinweisung auf ihre Kulturfachigkeit fuer Mittel-Europa beschrieben, *etc.* Wien und Ollmuetz, 18[58–]62. fol.

Krahe (—.).

Die Korbweiden-kultur. Aachen, 1879. 8°.

Kratzmann (Eduard).

De coniferis usitatis. Pragae, 1835. 8°.

Kraus (Karl).

Die Krankheiten der Hopfenpflanzen. Vortrag. Nuernberg, 1880. 8°.

KRAUSE (Johann Wilhelm).

Abbildungen und Beschreibung aller bis jetzt bekannten Getreidearten. Leipzig, 1835–37. fol.

Das Getraidebuch oder neuste Wanderungen durch das wissen-
schaftliche Gebiet der Getraide. Leipzig, 1840. 8°.

KRAUSE (Rudolf Wilhelm).

De Cardamomis. Jenae, 1704. 4°.

KRAUSS (J. C.).

Afbeeldingen der artsseny-gewassen met derzelver nederduitsche
en latynsche beschryyvingen. Amsterdam, 1796–1800. 4 vols.
8°. [*Anon.*]

KREBS (F. L.).

Vollstaendige Beschreibung und Abbildung der saemmtlichen
Holzarten, welche in mittleren und noerdlichen Deutschland
wild wachsen. Braunschweig, 1827–35. fol.

KRUEDENER (*Baron* C.).

Der Flachsbau mit Berucksichtung der Verhaeltnisse in Oesterreich. Wien, 1874. 8°.

KREUTZER (Karl Joseph).

Oestereichs Giftgewaechse. Wien, 1838. 8°.

Beschreibung und Abbildung saemmtlicher essbaren Schwaemme,
deren Verkauf auf den niederoesterreichschen Maerkten gezetz-
lich gestattet ist. Wien, 1839. 8°.

KROENISHFRANCK (*pseud.* [*i.e.* Arthur ÉLOFFE]).

Guide pour reconnaitre les champignons comestibles et vénéneux
du pays de France. Paris, [1869]. 16°. [Ed. 2, title varied,
1880.]

KROMBHOLZ (Julius Vincenz von).

Conspectus fungorum esculentorum, qui per decursum anni 1820
Pragae publica vendebantur. Programm. Prag, 1821. 8°.

Naturgetreue Abbildungen und Beschreibungen der essbaren,
schaedlichen und verdaechtigen Schwaemme. Prag, 1831–47.
fol.

KUBINYI (A.).

Plantae venenosae Hungariae. Mérges növényik, *etc.* Budae,
1842. 8°.

KURZ (Sulpiz).

The Forest Flora of British Burma. Calcutta, 1877. 2 vols. 8°.

KUNTZE (Otto).

Cinchona.   Arten, Hybriden, und Cultur der Chinabaeume.
Monographische Studien nach eigenen Beobachtungen in den
Anpflanzungen auf Java und im Himalaya. Leipzig, 1878. 8°.

L., (J. B.).

Principal economic products from the Vegetable Kingdom. Arranged under their respective Natural Orders, with the Names of the Plants, and the parts used in each case. London, 1872. 8°.

LABORIE (P. J.).

The Coffee Planter of St. Domingo. London, 1798. 8°.

An abridgment of the Coffee Planter of St. Domingo . . . also notes on the cultivation of the medicinal Cinchonas, by W. G. McIvor. Madras, 1863. 8°.

LACHAUME (—.).

The Cave Mushroom. [Extract.] *Pharm. Journ.* III. viii. (1878) 568–569.

LACIPIÈRE (P. Léopold).

Des émétiques végétaux en général, et des émétiques indigènes en particuliére. Strasbourg, 1854. 8°.

LADUREAU (A.).

Troisiéme mémoire sur la culture de la betterave à sucre. Lille, 1877. 8°.

Culture du lin. Étude sur les causes des maladies du lin. Lille, 1878. 8°.

Études sur la culture de la betterave à sucre (1878). Lille, 1880. 8°.

Note sur la luzerne du Chili (Medicago apiculata) et son utilisation agricole. Lille, 1880. 8°.

LAFFEMAS (B. de).

La façon de faire et semer la graine de menrier, les élever en pépinières et les replanter aux champs, gouverner et nourrir les vers à soye au climat de la France, plus facilement que par les mémoires de tous ceux qui en ont escript. Nouvelle édition, revue sur celle de Pierre Pautonnier, Paris, 1604. Montpellier, 1877. 8°.

LAGASCA (Mariano).

Memoria sobre las plantas barrilleras de España. Madrid, 1817. 8°.

LAGERGREN (Joh.).

Svenska matvaexters insamling och foervaring. Stockholm, 1880. 12°.

LAGUNA (Maximo).

Resúmen de los trabajos verificados por la comision de la flora forestal española durante los años de 1867 y 1868. Madrid, 1870. 4°.

Coniferas y amentáceas españolas. Madrid, 1878. 4°.

LAIRD (W.).

On the adulteration of Annatto. *Pharm. Journ.* II. x. (1868) 156–157.

LAMBERT (Aylmer Bourke).

A description of the genus Cinchona, comprehending the various species of vegetables from which the Peruvian and other barks of a similar quality are taken. Illustrated by figures of all the species hitherto discovered. To which is prefixed Prof. Vahl's Dissertation on this genus, *etc.* London, 1797. 4°.

A Description of the genus Pinus, illustrated with figures, directions relative to the cultivation, and remarks on the use of the several species. London, 1803–24. 2 vols. fol. Ed. 2. 3 vols. 1828–37. Also in 8°. 1832.

An illustration of the genus Cinchona, comprising descriptions of all the officinal Peruvian barks, including several new species, Baron de Humboldt's account of the Chinchona forests of South America; and Laubert's memoir on the different species of Quinquina. To which are added, several dissertations of Don Hippolito Ruiz, on various medicinal plants of South America. London, 1821. 8°.

LAMBERT (Ernest).

Exploitation des forêts de chêne-liège, et des bois d'olivier en Algérie. Paris, 1860. 8°.

Eucalyptus. Culture exploitation et produit, son role en Algérie. Paris, 1873. 8°. Ed. 2. 1874.

LAMPRECHT (Conrad).

Der Hopfen. Inaugural-Dissertation. Breslau, 1874. 8°.

LANDERER (Xaverias).

On Senna. [From the *Repert. fuer die Pharm.*] *Pharm. Journ.* v. (1845) 84.

On oriental copal. [From the *Repert. fuer die Pharm.*] *Pharm. Journ.* v. (1845) 186.

On Salep. *Pharm. Journ.* ix. (1850) 435–436.

Labdanum creticum. *Pharm. Journ.* x. (1851) 349–350.

On Cyprian Labdanum. Translated and communicated by Daniel Hanbury. *Pharm. Journ.* xi. (1851) 6–7.

On the poisonous plants of Greece. [From the *Central Blatt.*] *Pharm. Journ.* xxi. (1851) 120–121.

On Cassia Fistula. *Pharm. Journ.* xi. (1851) 201–202.

On Cornus mascula [=C. Mas, *Linn.*]. *Pharm. Journ.* xii. (1852) 63–64.

On the varieties of Manna not produced by the Ash. *Pharm. Journ.* xiii. (1854) 411–412.

On the use of common Rue, and of Peganum Harmala in Greece. [Extract from a letter.] *Pharm. Journ.* xiv. (1855) 369.

LANDERER (Xaverias) *continued :*—
Pharmaceutical notes on the Cornelian Cherry. [Cornus Mas, *Linn.* From *The Dublin Hospital Gazette.*] *Pharm. Journ·* II. 'i. (1860) 431.

LANDOLT (El.).
Die forstlichen Zustande in den Alpen und im Jura, *etc.* Bern, 1863. 8°.

Der Wald, seine Verjuenging, Pflege und Benutzung. Bearbeitet fuer das Schweizervolk. Zuerich, 1866. 8°. Ed. 2. 1871.

Der Wald in Haushalt der Natur und der Menschen, *etc.* Zuerich, 1870. 8°.

LANESSAN (J. L. de).
Gamboge [Abstract]. *Pharm. Journ.* III. ii. (1872) 848–849.

Manuel d'histoire naturelle médicale. Deuxième partie : étude des plantes phanérogames médicinales, suivi e d'un tableau des médicaments d'origine végétale, *etc.* Paris, 1879. 18°.

LANG (John Dunmore).
Queensland, Australia ; . . . the future cotton-field of Great Britain, *etc.* London, 1861. 8°.

LANGE (Johan Martin Christian).
Om de Sygdomme hos vore vigtigste dyrkede Planter, som Fremkaldes ved Rustsvampe, snyltende paa forskjellige Vaert-planter og om Midlerne til at indskraenke deres Udbredelse. Kjoebenhavn, 1880. 8°.

—— Om rostsjukdomar hos våra vigtigaste odlade vaexter och om medlen att inskräenka deras spridning. [Ed. by Ch. Jacobson.] Stockholm, 1880. 8°.

LANGE (Johan Martin Christian), & E. ROSTRUP.
De danske Foderurter. Udarbejdet paa Grundlag af S. Drejers, ' Anvisning til at kjende de danske Foderurter' tredje Udgave, *etc.* Kjoebenhavn, 1877. 8°.

LANGETHAL (Christian Eduard).
Lehrbuch der landwirthschaftlichen Pflanzenkunde fuer praktische Landwirthe und Freunde des Pflanzenreiches. i. Suessgraeser. ii. Klee- und Wickpflanzen. iii. Hackfruechte Handelsge-waechse und Kuechenkraeuter. Jena, 1841–45. 3 vols. 8°. Ed. 2. 1847–53. Ed. 3. 1855–64. Ed. 4 (vol. i.). 1866. Ed. 5 (Handbuch der landwirthschaftlichen Pflanzenkunde und des Pflanzenbaues). Berlin, 1874–76. 4 vols.

> i. Gras und Getreide. ii. Klee- und Wickpflanzen. iii. Hack-fruechte Handelsgewaechse, Gemuese und Apothekerkraeuter. iv. Der Obstbau, der Beerenbau, und die wildwachsenden Holzarten in Bereiche der Landwirthschaft.

LANGETHAL (Christian Eduard) *continued :*—

    Beschreibung der Gewaechse Deutschlands nach ihren natuer-
lichen Familien und ihrer Bedeutung fuer die Landwirthschaft.
Jena, 1858. 8°. Ed. 2. 1868.

LANKESTER (Edwin).

    Lecture on the structure, affinities, and medical properties of the
Natural Order Ranunculaceae. *Pharm. Journ.* ii. (1842)
198–203.

    Report of lectures on the natural history of plants yielding food,
etc. London, 1845. 12°.

LA ROQUE (—.)

    Gruendliche und sichere Nachricht vom Cafee- und Cafee-Baum,
etc. Leipzig, 1717. 4°.

LARSHOFF (—.).

    En kort Afhandling for Landmaend om Hoer-Silberedelse, *etc.*
Kjoebenhavn, 1872. 8°.

LA SAGRA (Ramon de).

    Description et culture de l'ortie de le Chine, précédé d'une notice
sur les diverses plantes qui portent ce nom, *etc.* Paris, [1870].
12°.

LASCELLES (Arthur R. W.).

    A treatise on the nature and cultivation of Coffee, *etc.* London,
1865. 8°.

LASLETT (Thomas).

    Timber and Timber Trees, native and foreign. London, 1875. 8°.

LASTEYRIE DU SAILLANT (Charles Philibert de).

    Du pastel, de l'indigotier, et des autres végétaux dont ou peut
extraire une couleur bleue ; avec une instruction detaillée sur
la culture et la préparation du pastel. Paris, 1811. 8°.

    —— Del guado e di altri vegetabili da cui si può estrarre un color
turchino, colla descrizione della coltura del guado, della pre-
parazione del pastillo, *etc.* Roma, 1811. 8°.

    Du cotonnier et de sa culture, ou traité sur les diverses espèces
de cotonniers ; sur la possibilité et les moyens d'acclimater
cet arbuste en France ; sur sa culture dans différens pays,
principalement dans le midi de l'Europe, et sur les propriétés
et les avantages économiques, et les économiques industriels et
commerciaux du coton. Paris, 1808. 8°.

LATTRE (—. de).

    Comité agricole de l'arrondissement de Dieppe. Notions théoriques
et pratiques sur la culture du lin, et sur celles du colza et du
tabac. Dieppe, 1864. 8°.

LAUBERT (Charles Jean).

Recherches botaniques, chimiques, pharmaceutiques sur le quinquina. Paris, 1816. 8°.

LAUCHE (W.).

Deutsche Dendrologie. Systematische Uebersicht, Beschreibung, Culturanweisung und Verwendung der in Deutschland mit oder ohne Decke aushaltenden Gehoelze. Berlin, 1880. 8°.

LAVALETTE (A. de).

Culture et ensilage du maïs. [Extract.] Paris, 1877. 8°.

LAVALLE (Jean).

Traité pratique des champignons comestibles, . . . leurs propriétés alimentaires, leur culture, *etc.* Paris, 1852. 8°.

LAVALLÉE (Alphonse).

Les nouveaux conifères du Colorado et de la Californie. Paris, [1875]. 8°.

L'origine de la pomme de terre et son introduction en Europe. [Extract.] Paris, 1877. 8°.

Arboretum Segrezianum. Enumération des arbres et arbrisseaux cultivés à Segrez (Seine-et-Oise) comprenant leur synonymie et leur origine, avec l'indication d'ouvrages dans lesquels ils se trouvent figurés. Paris, 1877. 8°.

Arboretum Segrezianum. Icones selectae arborum et fruticum in hortis Segrezianis collectorum. Description et figures des espèces nouvelles, rares ou critiques de l'arboretum de Segrez. Paris, 1880, *etc.* fol.—>

LAWES (John Bennett), J. H. GILBERT, & Maxwell T. MASTERS.

Agricultural, botanical, and chemical results of experiments on the mixed herbage of permanent meadow, conducted for more than twenty years in succession on the same land. Part I. *Phil. Trans.* clxxi. (1880) pp. 289–416. Part II. (Botanical results.) *Proc. Royal Soc.* xxx. (1880) pp. 556–557.

> The details of Messrs. Lawes and Gilbert's experiments will be found in a series of papers published in the Journal of the Royal Agricultural Society.

LAWSON (Peter).

The agriculturist's manual, being a familiar description of the agricultural plants cultivated in Europe. Edinburgh, 1836. 8°.

Pinetum britannicum, containing a descriptive account of all hardy trees of the Pine tribe cultivated in Great Britain. Edinburgh, 1866, *etc.* fol.—>

LAWSON (Peter, & Charles).

Synopsis of the vegetable products of Scotland, forming a
descriptive account of the collection exhibited at the Great
Exhibition of the industry of all nations. Edinburgh, 1851. 4°.

Agrostographia, a treatise on the cultivated grasses and other
herbage and forage plants. Ed. 4. Edinburgh, 1853. 4°. Ed.
6. (by D. Syme) 1877.

LEARED (Arthur).

On the use of the Leaves of Globularia Alyssum as a purgative.
[From *Med. Review.*] *Pharm. Journ.* II. ii. (1861) 431-432.

Notes on some drugs collected in Morocco. *Pharm. Journ.* III.
iii. (1873) 621-625. *See also* pp. 638-639.

LEARED (Arthur), & Edward Morell HOLMES.

Notes on Morocco drugs. *Pharm. Journ.* III. v. (1874) 521-523.

Further notes on Morocco drugs. *Pharm. Journ.* III. vi. (1875)
141-142.

LEAVITT (O. S.).

The culture and manufacture of flax and hemp. *Report, Commissioner of Patents; Agriculture.* Washington, (1862) pp. 83-118.

LEBEAUD (—.).

Manuel de l'herboriste, ou description succincte des plantes
usuelles indigènes, *etc.* Paris, 1825. 12°.

LEBEUF (Valentin Ferdinand).

Culture des champignons de couches, et de bois, et de la truffe,
ou moyen de les multiplier, reproduire, accomoder. Paris,
1878. 18°.

LEBLOND (Jean Baptiste).

Essai sur l'art de l'indigotier, pour servir à un ouvrage plus
étendu, lu et approuvé par l'Academie des sciences. [Paris],
1791. 8°.

Observations sur le cannelier de la Guyane française. Cayenne,
1795. 8°.

Memoire sur la culture du cotonnier dans les terres basses, dites
Palétuviers à la Guyane française. Cayenne, 1801. 4°.

LECOCQ (Henri).

Traité des plantes fourragères, ou flore des prairies naturelles et
artificielles de la France et de l'Europe centrale; ouvrage
contenant la description, les usages et qualités de toutes les
plantes herbacées ou ligneuses qui peuvent servir à la nourriture des animaux, *etc.* Ed. 2. Paris, 1862. 8°.

LE COMTE (C. E. A.).

Culture et production du café dans les colonies. Paris, 1865. 8°.

LECOUTEUX (Édouard).

Culture et ensilage du maïs-fourrage, et des autres fourrages verts. Paris, 1875. 18°.

LE DOUX (Chr.).

Note sur la culture de divers végétaux et particulièrement du panais fourrager, dans la Lozère. [Extract.] Paris, 1877. 8°.

LE DUC (William G.).

Letter . . . to Hon. J. W. Johnson . . . on Sorghum sugar. Washington, 1880. 8°.

LEENHARDT (—.).

Conférence sur la culture de la garance. Avignon, 1875. 8°.

LEES (William Nassau).

Tea and Cotton cultivation in India. London, 1863. 8°. *

Another word on Tea cultivation in Eastern Bengal. Calcutta, 1867. 8°.

LEFRANC (Édouard).

Étude botanique, chimique et toxicologique sur l'Atractylis gummifera (el heddad des Arabes). Paris, 1866. 8°.

Des Chamaeléons noir et blanc des anciens, Cardopatium orientale, *Spach*; et Atractylis gummifera, *L.* Botanique et matière medicale. Paris, 1867. 8°.

LEGER (A.).

La ramie et son exploitation industrielle. Lyon, 1881. 8°.

LEHNERT (Hugo).

Der Hopfenbau. Practische Anleitung zum richtigen Betriebe desselben. Berlin, 1877. 8°.

LEHR (Georg Philipp).

De Olea europaea. D. botanico-medica. Goettingae, 1779. 4°.

LEINGRE (—.).

Notice sur l'Eucalyptus globulus. Paris, 1875. 8°.

LÉMERY (Nicolas).

Dictionnaire ou traité universel des drogues simples, mis en ordre alphabétique. Paris, 1698. 4°. Ed. 2. 1714. Ed. [3.] by S. Morelot, 1807. 2 vols. 8°.

—— Vollstaendiges Materialen-Lexicon, verdeutscht durch Chr. Fr. Richter. Leipzig, 1721. fol.

LEMOINE (Constant), & —. CHARBONNEAU.

Abrégé du cours d'arboriculture théorique et pratique, par demandes et par réponses, à l'usage des jardiniers, propriétaires et cultivateurs. Angers, 1872. 18°.

Abrégé d'arboriculture théorique et pratique, redigè par demandes et par réponses. Rennes, 1880. 16°.

LENGLEN (Charles).

La culture de la betterave dans ses rapports avec la sucrerie indigène. Arras, 1876. 8°.

LENZ (Harold Othmar).

Die nuetzlichen, schaedlichen und verdaechtigen Schwaemme, *etc.* Gotha, 1831. 8°. Atlas 4°. Ed. 2. 1840. Ed. 3. 1862. Ed. 4. 1868. Ed. 5. [by A. Roese,] 1874.

LEO (Julius).

Taschenbuch der Arzneipflanzen, oder Beschreibung und Abbildung saemmtlicher offizinellen Gewaechse. Berlin, 1826–27. 4 vols. 8°.

LEON (John A.).

The art of manufacturing and refining Sugar, *etc.* London, 1850. fol.

LEPAGE (Richard C.).

Papers on the plant Gynocardia odorata, from which the Chaulmoogra oil is obtained. London, 1878. 8°.

LEPAIRE (L.).

De la culture de la betterave, des soins à lui donner pour augmenter le poids de la récolte, conseils aux cultivateurs. Chalon-sur-Saône, 1881. 8°.

LEQUAIN (Fr.).

La coltivazione del rizo, con provvedimenti di salubrità. Torino, 1878. 16°.

LEROY (Alphonse Vincent Louis Antoine).

Mémoire sur le Kinkina français. Paris, 1808. 8°.

LEROY (André).

Dictionnaire de pomologie, contenant l'histoire, la description, la figure des fruits anciens et des fruits modernes les plus généralement connues et cultivés. Angers, 1869. 8°.

LESCHENAULT DE LA TOUR (Louis Théodore).

Notice sur le cannelier de l'île de Ceylon, sur sa culture et ses produits. A St. Denis, île Bourbon, 1821. 4°.

LESCHER (F. H.).

On Silphium, or Asafoetida. *Pharm. Journ.* II. ix. (1868) 588–593.

LÉTELIÉ (J. A.).

Le cotonnier, son histoire, sa culture et ses succès en Algérie, *etc.* Marennes, 1863. 8°.

LETELLIER (Jean Baptiste Louis).

Dissertation sur les propriétés alimentaires, médicales et vénéneuses des champignons, qui croissent aux environs de Paris. D. Paris, 1826. 4°.

Histoire et description des champignons alimentaires et vénéneux qui croissent aux environs de Paris. Paris, 1826. 8°.

LETTSOM (John Coakley).

The natural history of the Tea-tree, with observations on the medical qualities of Tea and effects of Tea-drinking. London, 1772. 4°. Ed. 2. 1799.

LEVEILLÉ (Joseph Henri).

Notice sur le genre Agaric, consideré sous les rapports botanique, économique, médical et toxologique. Paris, 1840. 8°.

LEWIS (G. C.).

Coffee planting in Ceylon; past and present. Colombo, 1855. 8°.

LIAUDET (Philipp).

Memoranda der praktischen Botanik in ihrer Anwendung auf Materia medica. Weimar, 1851. 8°.

LIEBICH (Christian).

Compendium der Forstwissenschaft. Wien, 1854. 8°.

LIEBIG (Justus, *Baron* von).

Coffee. [From *Popular Science Review*.] *Pharm. Journ.* II. vii. (1866) 412–416.

LIGNAC (L.).

Dicotylédones. Caractères des principales familles et plantes etudiées en médicine. Leurs usages thérapeutiques. Paris, 1879. 8°.

Monocotylédones et acotylédones. Principales familles, *etc.* Paris, 1879. 8°.

LIGON (Richard).

A True and exact history of Barbados, *etc.*, [with the process of sugar-making.] London, 1657. fol. (Another edition, 1673.)

LINDLEY (John).

Flora medica: a botanical account of all the more important plants used in medicine, in different parts of the world. London, 1838. 8°.

Medical and economical Botany. London, [1849]. 8°. Ed. 2. 1856.

The Sassy tree of Western Africa. [Correction of specific name, given by PROCTER.] *Pharm. Journ.* xvi. (1857) 373.

LINDLEY (John), & Thomas MOORE.

Treasury of Botany. London, 1866. 2 vols. 8°. Ed. [2.] 1874.
    The title on the cover reads, "Maunder's Treasury of Botany."

LINDSAY (William Lauder).

The Toot-poison of New Zealand. [*Proc. Brit. Assoc.*] *Pharm. Journ.* II. v. (1864) 371–373.

On the conservation of forests in New Zealand. *Journ. Bot.* vi. (1868) 38–46.

LINK (Heinrich Friedrich).

Handbuch zur Erkennung der nutzbarsten und am haeufigsten vorkommenden Gewaechse. Berlin, 1829–33. 3 vols. 8°.

LINKE (J. R.).

Lehrbuch der medicinisch-pharmaceutischen Pflanzenkunde fuer Aerzte, Apotheker, Droguisten, *etc.* Leipzig, 1863. 8°.

Atlas der Giftpflanzen oder Abbildung und Beschreibung der den Menschen und Thieren schaedlichen Pflanzen. Ed. 2. Leipzig, 1867–68. 4°. Ed. 3. 1875.

LISCHWITZ (Johann Christoph).

Plantae diureticae cum habitu externo tum quoque charactère botanico diversae, charactere autem pharmaceutico congeneres usuque eaedem. D. Kilonii, 1739. 4°.

Dissertatio, sistens plantas anthelminticas et habitu externo et toto genere botanico diversas, charactere autem pharmaceutico usuque medicinali congeneres. Kilonii, 1742. 4°.

LIONS (A.).

Essai sur les végétaux utiles qui croissent spontanément dans le département des Bouches-du-Rhône, qui y sont cultivés ou qui seraient susceptibles de l'etre, *etc.* Marseille, 1864. 8°.

LIOTARD (L.).

Memorandum on (Vegetable) materials in India suitable for the manufacture of paper. Calcutta, 1880. fol.

LLOYD (J. U.).

Anemopsis californica (Hooker). Yerba mansa. [*Amer. Journ. Pharm.*] *Pharm. Journ.* III. x. (1880) 666–667.

LOARER (E.).

Thé d'Himalaya, ses productions naturelles. Culture du thé dans l'Inde. Paris, 1868. 8°. *

LOCKHART (William).

Ginseng. [Extract from Lockhart's *Medical Missionary in China.*] *Pharm. Journ.* II. iii. (1861) 332–333.

LOCK (Charles George Warnford).

Notes on some neglected fibres. *Journ. Soc. Arts,* xxviii. (1880), 912–914, 916–918.

"Mhowa" or "Mahwah," an Indian food-tree. *Journ. Soc. Arts,* xxix. (1881), 285.

Rose Oil or Otto of Roses. *Journ. Soc. Arts,* xxix. (1881), 237–239. Extract in *Pharm. Journ.* III. xi. (1881), 899–900.

LOEBE (William).

Die Handelspflanzen, Wurzel-, Knollen-, Kuechengewaechse und essbaren Schwaemme. Leipzig, 1862–3. 4°. Ed. 2. 1868. Ed. "2." 1872.

LOEBE (William) *continued* :—

Die Getreidearten und Huelsenfruchte. Leipzig, 1862. 4°. [Ed.
2.] 1868–72.

Die Graeser der Wiese und des Waldes. Leipzig, 1863. 4°.
Ed. 3. 1874–75.

Die Unkraeuter des Feldes und des Waldes. Leipzig, 1863. 4°.
Ed. 2. 1870. Ed. "2." 1872.

Die Futterkraeuter. [Ed. 2. by David Dietrich.] Leipzig,
1863. 4°. Ed. 3. 1869. Ed. "3." 1871–72.

Landwirthschaftliche Flora Deutschlands oder Abbildung und
Beschreibung aller fuer Land- und Hauswirthe wichtigen
Pflanzen. Ed. 2. Leipzig, 1862. 4°. Ed. 3. 1868–72.
Ed. 4. Dresden, 1876.

Die Krankheiten der Kultur-Pflanzen auf Aeckern, in Obstan-
lagen, Wein-, Gemuese-, und Blumengaerten, Anleitung zur
Erkenntniss, Verhuetung und Heilung aller innerlichen und
aeusserlichen Krankheiten des Getreides, der Huelsenfruechtes
Futterpflanzen, *etc.* Hamburg, 1864. 8°.

Anleitung zum rationellen Anbau der Handelgewaechse; der
Fabrik-, Farbe-, Gewuerz-, . . . Pflanzen behufs Erziehung
einer hoeheren Bodenrente. Stuttgart, 1867. 8°. Hannover,
1878–70.

> Also separately, in seven parts: 1. Gewuerzpflanzen. 2. Fabrik-
> pflanzen. 3. Gespinnstpflanzen. 4. Oelpflanzen. 5. Farbepflanzen.
> 6. Arznei- und Speizereipflanzen. 7. Feldgaertnerei.

LOEFFELHOLZ-COLBERG (Friedrich, *Freiherr* von).

Die Bedeutung und Wichtigkeit des Waldes, Ursachen und
Folgen der Entwaldung, die Wiederbewaldung, mit Ruecksicht auf Pflanzenphysiologie, Klimatologie, . . . aller Laender
fuer Forst- und Landwirthe, Nationaloekonomen und alle
Freunde des Waldes aus der einschlag. Literatur systematisch
und Kritisch, nachegewiesen und bearbeitet. Leipzig, 1872. 8°.

LOEFFLER (Karl).

Anbau und Ausbeute der Industriegewaechse. Fuer deutsche
Landwirthe. Wittenberg, 1863. 8°.

> I. Die Cichorie. II. Die schwarze Malve. III. Der Krapf.

LOEUILLART-D'AVRIGNI (A. E. C.).

Principes de botanique médicale. Paris, 1821. 12°.

LOHREN (A.).

Deutschlands Flachsbau. Eine im deutschen Reichstage 1879
unerledigt gebliebene Zollposition. Berlin, 1880. 8°.

Loiseleur-Deslongchamps (Jean Louis Auguste).

Manuel des plantes usuelles indigénes, ou histoire abregée des plantes de France, distribuées d'après une nouvelle méthode. Paris, 1819. 2 vols. 8°.

Lorenz (Josef R.).

Die Bodencultur auf der Wiener Weltausstellung, 1873, *etc.* Wien, 1874. 3 vols. 8°.

Loring (Frederick W.), & C. F. Atkinson.

Cotton Culture and the South considered with reference to Emigration. Boston [U.S.], 1869. 8°.

Loudon (John Claudius).

Hortus britannicus : a catalogue of all the plants indigenous, cultivated in, or introduced into Britain. With Supplement. London, 1830. 8°. Ed. 2. 1832. Ed. 3. with second supplement, 1839.

Hortus lignosus Londinensis, or a catalogue of all the ligneous plants, indigenous and foreign, hardy and half-hardy, cultivated . . . in the neighbourhood of London. London, 1838. 8°.

Arboretum et fruticetum britannicum, or the trees and shrubs of Britain, native and foreign, hardy and half-hardy, pictorially and botanically delineated, and scientifically and popularly described; with the propagation, culture, management and uses in the arts, in useful and ornamental plantations, and in landscape gardening. Preceded by a historical and geographical outline of the trees and shrubs of temperate climates throughout the world. London, 1838. 8 vols. 8°.

An encyclopaedia of trees and shrubs; being the Arboretum . . . abridged, *etc.* London, 1842. 8°.

The Derby Arboretum ; containing a catalogue of the trees and shrubs included in it, *etc.* London, 1840. 8°.

Lovén (F. A.).

Om parasitsvampar och deras inflytande på skogskulturen. Akademisk afhandling. Lund, 1874. 8°.

Low (James).

A Dissertation on the Soil and Agriculture of the British Settlement of Penang, or Prince of Wales Island, *etc.* Singapore, 1836. 8°.

Low (Hugh).

Sarawak ; its inhabitants and productions : being notes during a residence in that country with H. H. the Rajah Brooke. London, 1848. 8°.

LUDLOW (E.).
> Memorandum on Cardamom cultivation in Coorg. Bangalore,
> 1869. 8°.

LUDOVICI (Leopold).
> Rice cultivation; its past history and present condition, with
> suggestions for its improvement. Colombo, 1867. 8°.

LUDWIG (Christian Friedrich).
> Handbuch der Botanik. Zu Vorlesungen fuer Aerzte und Oeko-
> nomen. Leipzig, 1800. 8°.

LUEDERS (Philipp Ernst).
> Umstaendliche Beschreibung vom dem Leinbau, *etc.* Flensburg,
> 1760. 8°.
> Der Lein-Bau, in seiner verbesserten Gestalt. Flensburg, 1765. 8°.
> Kurze Anleitung zum Leinbau. Flensburg, 1770. 8°.
> Kurze Anleitung zum Hopfenbau. Flensburg, 1782. 8°.
> Von dem Hopfen-Bau. Gluecksburg, 1786. 8°.

LUEDICKE (A.).
> Ueber die Papierfabrikation in Japan. (*Jahresb. Ver. Naturw.
> Braunschw.* 1879–80, pp. 81–89.)

LUERSSEN (Christoph).
> Medicinisch-pharmaceutische Botanik. Handbuch der systemat-
> ischen Botanik fuer Botaniker, Aerzte und Apotheker. Leipzig,
> 1876, *etc.* 8°.→

LUKMANOFF (Athanase de).
> Nomenclature et iconographie des canneliers et camphriers.
> Paris, 1878. 4°.

LUMSDAINE (*Dr. —*).
> Cultivation of nutmegs and cloves in Bencoolen. [Published in
> 1820.] *Pharm. Journ.* xi. (1852) 516–520.

LUNDAHL (B.).
> Tabak ist Gift! Physischer und psychischer Einfluss des Tabaks
> auf den menschlichen Organismus . . . aus dem Schwedischen.
> Berlin, 1866. 8°.

LYMAN (Joseph B.).
> Cotton Culture. With an additional chapter on cotton seed and
> its uses by J. C. Sypher. New York, [1868]. 8°.

LYON (P.).
> A treatise on the physiology and pathology of trees, *etc.* Edin-
> burgh, 1816. 8°.

MACCHIATI (Luigi).
> Notizie utili sugli alberi e gli arbusti della Sardegna. Sassari,
> 1877. 8°.

MacCLELLAND (John).

Report on the physical condition of the Assam Tea plant, with reference to geological structure, soils, and climate. [Calcutta, 1838.] 8°.

Papers relating to the measures adopted for introducing the cultivation of the Tea plant in India. Calcutta, 1839. fol. [*Anon.*]

MACEDO (A. de).

Notice sur le Palmier Carnauba. Paris, 1867. 8°.

MACGOWAN (D. J.).

The Tallow tree and its uses. [*Scientific American.*] *Pharm. Journ.* III. ii. (1872) 1034.

MacHENRY (G.).

The Cotton trade; its bearing upon the prosperity of Great Britain and commerce of the American Republics, *etc.* London, 1863. 8°.

The Cotton question. London, [1864]. 8°. Revised ed. [1864].

McIVOR (William Graham).

Notes on the propagation and cultivation of the medicinal Chinchonas or Peruvian bark trees. Madras, 1863. 8°.

Analysis of Chinchona bark and leaves. [Letter from W. G. McIvor, accompanying specimens, and report thereon by J. E. Howard.] *Pharm. Journ.* II. v. (1863) 367–369.

Analysis of Chinchona bark and leaves, received June 21st, 1865. [Letter from W. G. McIvor, with specimens, and report by J. E. Howard.] *Pharm. Journ.* II. vii. (1866) 419–421.

MACKAY (John).

Commercial otto or attar of Roses. *Pharm. Journ.* xviii. (1859) 413.

MACLAGAN (J. M'Grigor).

On the Bebeeru tree [Greenheart] of British Guiana. [From *Trans. Royal Soc. Edinb.*] *Pharm. Journ.* iii. (1843) 177.

Gutta Percha, a peculiar variety of caoutchouc. [From the *Edinb. Phil. Journ. and Gaz.*] *Pharm. Journ.* v. (1846) 472–473.

On the manner of the propagation of Colchicum autumnale. [From the *Journ. Medical Science.*] *Pharm. Journ.* xi. (1852) 416–418.

MACMILLAN (J. Laker).

Chrysophanic acid [and the source of Araroba]. *Pharm. Journ.* III. ix. (1879) 755–757.

McMurtrie (William).

Report on the culture of Sumac in Sicily, and its preparation for market in Europe and the United States. Washington, 1880. 8°.

McNab (James).

Notes on the propagation of the Ipecacuan plant (Cephaelis Ipecacuanha). *Trans. Bot. Soc. Edinb.* x. (1870) 318–324.

Notes on the 'Dogwood' of [gun-]powder manufacturers. *Trans. Bot. Soc. Edinb.* x. (1870) 324–327.

The cultivation of medicinal plants on railway banks. [Extract.] *Pharm. Journ.* III. v. (1874) 350.

Mac Rae (Alexander).

A manual of plantership in British Guiana, *etc.* London, 1856. 8°.

Madden (E.).

Observations on Himalayan Coniferae, being a supplement to the brief observations on some of the pines and other coniferous trees of the Northern Himalaya. Calcutta, 1850. 8°.

Madinier (Paul), & G. de Lacoste.

Guide du cultivateur du Sorgho à sucre, suivi de l'indication des diverses applications industrielles de cette plante, *etc.* Paris, [1856]. 8°.

Madiot (—.).

De la culture du Mûrier, reduite aux moyens les plus simples et les plus surs. Lyon et Paris, 1826. 8°.

Madriz (F. J.).

Cultivo del café ó sea manual teórico-prático sobre el beneficio de este frute con las mayores ventajas para al agricultor. Paris, 1869. 12°.

Magazines. See *Periodical Publications,* at end of Alphabetical List of Authors.

Maillot (Édouard).

Étude comparée du pignon et du ricin de l'Inde. Nancy, 1880. 8°.

Maisch (John M.).

The seeds of two species of Strychnos. [*Amer. Journ. Pharm.*] *Pharm. Journ.* III. ii. (1871) 23.

The so-called African saffron. [*Amer. Journ. Pharm.*] *Pharm. Journ.* III. ii. (1872) 824–825.

Pharmacognostical notes.—American indigenous plants. [*Amer. Pharm. Journ.*] *Pharm. Journ.* III. ii. (1872) 989–990.

Notes on some North American drugs. [*Amer. Journ. Pharm.*] *Pharm. Journ.* III. ii. (1874) 773–774.

MAISCH (John M.) *continued* :—

The constituents and properties of the genus Potentilla. [*Amer. Journ. Pharm.*] *Pharm. Journ.* III. v. (1875) 986–988.

Notes on the genus Teucrium. [*Amer. Journ. Pharm.*] *Pharm. Journ.* III. vii. (1876) 313.

The useful species of Viburnum. [*Amer. Journ. Pharm.*] *Pharm. Journ.* III. viii. (1878) 750–752.

MAISONNEUVE (Paul).

Étude sur la structure et les produits du camphrier de Bornéo ou Dryobalanops aromatica. Paris, 1876. 8°.

MAISTRE (Jules).

De l'influence des fôrets et des cultures sur le climat et sur le régime des sources. Montpellier, 1881. 8°.

MALAVOIS (—.).

De la culture de la canne et de la fabrication du sucre à l'Ile de la Réunion. Paris, 1861. 8°.

MALHERBE (Alfred).

Notice sur qulques espèces de chênes et spécialement du chêne liége (Quercus Suber). Metz, 1838. 8°.

MALLET (John William).

Cotton : the chemical, geological, and meteorological conditions involved in its successful cultivation in the Southern or Cotton States of North America. London, 1862. 8°.

MALTASS (Sidney H.).

On the production of Scammony in the neighbourhood of Smyrna. *Pharm. Journ.* xiii. (1853) 264–268.

On the production of opium in Asia Minor. *Pharm. Journ.* xiv. (1855) 395–400.

On Tragacanth and its adulteration. *Pharm. Journ.* xv. (1855) 18–21.

MALY (Joseph Karl).

Systematische Beschreibung der gebraeuchlichsten in Deutschland wildwachsenden oder Kultivirten Arzneigewaechse. Graetz, 1837. 8°.

Systematische Beschreibung der in Oesterreich wildwachsenden und Kultivirten Medizinalpflanzen. Wien, 1863. 8°.

Oekonomisch-technische Pflanzenkunde. Systematische Beschreibung der in der Garten- und Landwirthschaft, in Kuensten und Gewerben und in Forstwesen gebraeuchlichen kultivirten und wildwachsenden Pflanzen mit Angabe der Benutzung. Wien, 1864. 8°.

MANETTA (Filippo).

Il re cotone, ossia distretti cotoniferi del globo, considerati in relazione al loro clima; con la seconda edizione della guida per coltivare practicamente la pianta del cotone secondo il metodo Americano. Torino, 1863. 8°.

MANGIN (Arthur).

Les plantes utiles. Tours, 1869. 8°.

Les lichens utiles. [Extract.] Lyon, 1877. 8°.

Le Cacao et le chocolat considérés aux points de vue botanique, chimique, physiologique, agricole, commercial, industriel, et économique. Paris, 1860. 12°.

MANN (James A.).

The Cotton-trade of Great Britain, its rise, progress, and present extent, etc. London, 1860. 8°.

The Cotton-trade of India. A paper read before the Royal Asiatic Society. London, 1860. 8°.

Cocoa, its cultivation, manufacture and uses; its advantages and value as an article of food. London, 1860. 8°. Reprinted from Journ. Soc. Arts, viii. (1860) 775–780, 785–790, 795–800, 805–810.

MANN (Johann Gottlieb).

Deutschlands gefaehrlichste Giftpflanzen mit erlaeuterndem Texte. Stuttgart, 1829. fol.

Die auslaendischen Arzneipflanzen. Stuttgart, 1830–33. fol.

MANTEGAZZA (Paolo).

On the dietetic and medicinal properties of Erythroxylon Coca. [Extract] Pharm. Journ. II. i. (1860) 606–618.

MANTEUFFEL (Hans Ernst, Freiherr von).

Die Eiche, deren Anzucht und Abnutzung. Ein wohlmeinender Rathgeber fuer Eichenzuechter und Solche die es werden wollen. Ed. 2. Leipzig, 1874. 8°.

MANWARING (H. M.).

Treatise on the cultivation and growth of Hops. London, 1855, 12°.

MARAVIGNA (Carmelo).

Saggio di une Flora medica catanese, ossia catalogo delle principale piante medicinali che spontanamente crescono in Catania e ne' suoi contorni con la indicazione delle loro medichi azioni. Catania, 1829. 4°.

MARC (J.).

Sorghum halepense als Futterpflanze. (Oesterr. landw. Wochenblatt, 1880, p. 494.)

MARCHAND (Léon).

Botanique cryptogamique pharmaco-medicale. (Fasc. 1. Introduction à l'étude de cryptogames). Paris, 1880. 8°.

MARCOTTE (J.).

Étude générale du Matico. Historique, origine, étude botanique, chimique, pharmacologie, thérapeutique. Paris, 1863. 4°.

MARIEZ (Louis).

Manuel d'arboriculture et de viticulture théorique et pratique approprié aux départements du sud-ouest. Auch, 1873. 18°.

MARJOLLET (—.).

La cuscute et la décuscutage. Moutiers, 1878. 16°.

MARKHAM (Clements Robert).

Travels in Peru and India while superintending the collection of chinchona plants and seeds in South America, and their introduction into India. London, 1862. 8°.

On the introduction of the Cinchona plant into India. [*Trans. Med. and Phys. Soc. Bombay.*] *Pharm. Journ.* II. iii. (1862) 611–618. *Id.* iv. (1862) 27–34.

On the supply of Quinine and the cultivation of Chinchona plants in India. *Journ. Soc. Arts,* xi. (1863) 325–336.

On Chinchona cultivation in Ceylon. [Extract.] *Pharm. Journ.* II. vii. (1866) 521.

The Chinchona species of New Granada, containing the botanical description of the species examined by Drs. Mutis and Karsten; with some account of those botanists, and of the results of their labours. London, 1867. 8°.

The cultivation of Caoutchouc-yielding Trees in British India. *Journ. Soc. Arts,* xxiv. (1876) 475–482.

Peruvian Bark. A popular account of the introduction of Chinchona cultivation into British India. London, 1880. 8°.

MARKS (Hyman).

Botanical companion to the British Pharmacopoea. Dublin, 1873. 8°.

MARMÉ (W.).

Grundriss der Vorlesungen ueber Pharmacognosie des Pflanzen- und Thierreichs. Goettingen, 1880. 8°.

MAROLDA-PETILLI (Francesco).

Gli eucalitti; notizie raccolte. Roma, 1880. 8°.

MARQUART (Friedrich).

Beschreibung der in Maehren und Schlesien am haeufigsten vorkommenden essbaren und schaedlichen Schwaemme. Bruenn, 1842. 8°.

MARQUART (Friedrich) *continued :—*
  Die essbaren und schaedlichen Schwaemme beschrieben. All-
    muetz, 1856. 8°.
MARQUART (Ludwig Clamor).
  Die Scammoniumsorten des Handels, monographisch bearbeitet.
    Lemgo, 1836–37. 8°.
MARQUIS (Alexandre Louis).
  Essai sur l'histoire naturelle et médicale des Gentianes. Thèse.
    Paris, 1810. 4°.
MARSHALL (Henry).
  Contributions to the natural and economical history of the
    Cocoanut palm. London, 1832. 8°. Ed. 2. 1836.
MARSHALL (Humphrey).
  Arbustum americanum, the American grove, or an alphabetical
    catalogue of forest trees and shrubs, natives of the American
    United States. Philadelphia, 1785. 8°.
  —— Catalogue alphabétique, *etc.*, traduit, avec des observations
    sur la culture. Paris, 1788. 8°.
  —— Beschreibung der wildwachsenden Baeume und Stauden-
    gewaechse in den Vereinigten Staaten von Nordamerika, *etc.*
    Leipzig, 1788. 8°.
MARSHALL (L. T.).
  Hop Culture. *Report, Commissioner of Patents ; Agriculture.*
    Washington, 1862, pp. 289–293.
MARTIGNY (Julius).
  On a spurious yellow-bark. [From the *Pharm. Centr. Blatt.*]
    Pharm. Journ. vi. (1846) 232.
MARTIN (Robert Montgomery).
  The past and present state of the Tea trade of England, and of
    the continents of Europe and America, *etc.*   London, 1832.
    8°.
MARTINDALE (William).
  Jaborandi. *Pharm. Journ.* III. v. (1874) 364–366.
  Additional notes on Jaborandi, and its physiological action.
    Pharm. Journ· III. v. (1875) 561–562.
  Chian Turpentine. *Pharm. Journ.* III. x. (1880) 854–855. *Id·*
    xi. (1880) 271.
MARTINET (A.)
  L'elagage des essences forestières. Châteauroux, 1875. 8°.
  Notice sur l'élagage des arbres. (Exposition universelle de 1878.)
    Paris, 1878. 4°.

MARTINEZ RIBON (C.).

Nuevo metodo para el cultivo del cacao, *etc.* Ed. 2. Braine-le-
Comte, 1880. 16°.

MARTINI (—.).

Dissertatio epistolaris de oleo Wittnebiano seu Kajuput ab homine
Wolfenbuttelano in India orientali invento, in terras Bruns-
vicenses feliciter revocato ejusque saluberrimis effectis. [Wol-
fenbuettel], 1751. 4°.

MARTINY (Eduard).

Encyclopaedie der medizinisch- pharmaceutischen Naturalien- und
Rohwaarenkunde. Quedlinburg, 1843–54. 8°.

MARTIUS (Carl).

On wild senna. *Pharm. Journ.* xvi. (1857) 426–427.

MARTIUS (Carl Friedrich Philipp von).

Specimen materiae medicae brasiliensis, exhibens plantas medicinales, quas in itinere per Brasiliam observavit. [Monachii,]
1824. 4°.

Systema materiae medicae vegetabilis brasiliensis, *etc.* Lipsiae,
1843. 8°.

On Coca and Matico. [From *Gauger's Repert.*] *Pharm. Journ.*
ii. (1843) 660.

On the occurrence and geographical distribution of the genuine
Cinchona (Cinchona Condaminea) and of other Cinchona species
in the neighbourhood of Loxa. [From *Buchner's Repert.*]
Pharm. Journ. vii. (1847) 88–91, 184–191.

MARTIUS (Georg).

Pharmakologisch-medizinische Studien ueber den Hanf. D.
Erlangen, [1855]. 8°.

MARTIUS (Theodor Wilhelm Christian).

Abyssinian remedies for the cure of tape-worm. [From the
Jahrb. fuer prakt. Pharmacie.] *Pharm. Journ.* xi. (1851)
162–164.

On the Black Balsam of Peru. (In a letter to Dr. Pereira.)
[With Dr. Pereira's reply.] *Pharm. Journ.* xi. (1851) 202–205.

On Radix Uncomocomo (Aspidium athamanticum, *Kunze*).
A new remedy for tapeworm, from South Africa. *Pharm.
Journ.* xi. (1851) 261–262.

On Wai-fa, the unexpanded flower-buds of Sophora japonica, L.
[From the *Neues Jahrb. fuer Pharmacie.*] *Pharm. Journ.* xiv.
(1854) 64–65.

Culture of cochineal in the Canary Islands. *Pharm. Journ.* xiv.
(1855) 553–556.

MASON (Francis).

Materia medica, and pathology [in the Karen language]. Tavoy, 1848. 12°.

The Liquidamber tree of the Tenasserim Provinces. [From the *Journ. Asiatic Soc. Bengal.*] *Pharm. Journ.* viii. (1848) 243.

The Gum-Kino of the Tenasserim Provinces. [From the *Journ. Asiatic Soc. Bengal.*] *Pharm. Journ.* viii. (1849) 387–388.

The Natural Productions of Burmah ; or notes on the fauna, flora, and minerals of the Tenasserim provinces and the Burman Empire. Maulmain, 1850. 8°.

Burmah ; its people and natural productions, *etc.* Rangoon, 1860. 8°.

MASSA (Francesco).

L'oliva, dalla gemma al frantajo ; osservazioni pratiche. Lecce, 1880. 8°.

MASSE (J.).

Du traitement industriel des plantes filamenteuses qui peuvent être employées à la fabrication des tissues et du papier concurrement avec le coton, le lin et le chanvre. Lille, 1864. 8°.

MASSONNEAU (—.).

La betterave à sucre, sa culture et ses produits. Conseils aux interessés. Poissy, 1875. 18°.

MASTERS (J. W.).

A run through the Assam Tea gardens. Golaghat, 1863. fol.

MASTERS (Maxwell Tylden).

On the principles regulating the transfer of useful plants of one country to another. *Journ. Soc. Arts*, iv. (1856) 311–314.

MATHIEU (Auguste).

Description des bois, des essences forestières les plus importantes. Nancy, 1855. 8°.

Flore forestière. Description et histoire des végétaux ligneux qui croissent spontanément en France. Nancy, 1858. 8°. Ed 2. 1861. Ed. 3. 1877.

Les fôrets à l'exposition et au congrès international agricole et forestier de Vienne en 1873. Paris, 1875. 8°.

Le reboisement et le regazonnement des alpes. Paris, 1875. 4°.

MATTIOLI (—.), & —. PASQUALIGO.

Sulla fava del Calabar, discussione scientifica. Venezia, 1868. 8°.

MAUKE (Johann Gottlob).

Grasbuechlein, oder Anweisung, die schaedlichsten und nuetzlichsten inlaendischen Graeser kennen . . . zu lernen. Leipzig und Meissen, 1801. 4°. Ed. 2. Leipzig, 1818.

MAW (George).

The Sugar Maple. [From *The Gardeners' Chron.*] *Pharm. Journ.* III. ix. (1878) 186.

MAYER (Johann Christoph Andreas).

Abhandlung von dem Nutzen der systematischen Botanik in der Arznei- und Haushaltungskunst. Greifswald, 1772. 4°.

Einheimische Giftgewaechse, welche fuer Menschen am schaedlichsten sind. Berlin, 1798–1800. fol.

Vorzuegliche einheimische essbare Schwaemme. Anhang der Beschreibung der schaedlichen einheimischen Giftgewaechse. Berlin, 1801. fol.

MAYET (—.).

On starch, arrowroot, and sago. Extracted from a memoir of M. M.'s, by M. Guibourt. [From the *Journ. de Pharmacie.*] *Pharm. Journ.* vi. (1846) 37–40.

MECHOW (L.).

Die Cultur und Bewirthschaftung der Kiefern-Forstern fuer Forst- und Landwirthe. Osterburg, 1874. 8°.

MEDLICOTT (J. G.).

Cotton-handbook for Bengal, being a digest of information ... on the subject of the production of Cotton in the Bengal Provinces. Calcutta, 1862. 4°.

MEE (George).

Poisonous principle of bitter Cassava root. [*See* DANIELL.] *Pharm. Journ.* II. vi. (1864) 332–333.

MEINSHAUSEN (Karl).

Synopsis plantarum diaphoricarum florae ingricae, oder Notizen-Sammlung ueber die mannig-faltige Verwendung der Gewaechse Ingriens. St. Petersburg, 1869. 8°.

MEISNER (Leonard Ferdinand).

De caffe, chocolatae, herbae thee ac nicotianae, natura, usu et abusu anacrisis, medico-historico-diaetetica. Norimbergae, 1721. 12°.

MEITZEN (Hugo).

Ueber den Werthe der Asclepias Cornuti, Decne. als Gespinnstpflanze; D. Goettingen, 1862. 8°.

MELLER (Henry James).

Nicotiana; or the smoker's and snuff-taker's companion; containing the history of Tobacco; culture, medical qualities, *etc.* London, 1832. 12°.

MÉNE (Édouard).

Des usages du bambou en Chine. *Bull. Soc. imp. d'acclimatation,* Janvier, 1869.

MÉNE (Édouard) *continued* :—
Des produits végétaux de la Chine, et en particulier du bambou.
Loc. cit., Fevrier, 1869.

MÉNIER (Émile Justin).
Café, succedanés du café, cacao et chocolat, coca et thé matè.
Exposition universelle de 1867 à Paris. [Rapport du jury
international.] Paris, 1868. 8°.

MERREN (Daniel Carl Theodor).
Ueber den Cortex adstringens brasiliensis. Koeln, 1828. 8°.

METTENHEIMER (—.).
A spurious Rhatany root. [From the *Jahrb. prakt. Pharmacie.*]
Pharm. Journ. xi. (1852) 420.

METZ (—.), & Co.
Berichte ueber neuere Nutzpflanzen, insbesondere ueber die
Ergebnisse des Anbaues in verschiedenen Theilen Deutschlands.
Berlin, 1868. 8°.

METZ-NOBLAT (Alexandre de).
Le traitement des bois en France. Paris, 1881. 8°.

METZE (Wilhelm).
Ueber Baumanpflanzungen und deren besondern Nutzen fuer die
Marschlaender, als Mittel, die Deiche zu befestigen und
Deichbruecke zu verhueten. Schleswig, 1820. 8°.

METZGER (Johann).
Die Getraidearten und Wiesengraeser in botanischer und oecono-
mischer Hinsicht bearbeitet. Heidelberg, 1841. 8°.
Landwirthschaftliche Pflanzenkunde, oder praktische Anleitung
zur Kenntniss und zum Anbau der fuer Oekonomie und Handel
wichtigern Gewaechse. Heidelberg, 1841. 8°.

MEYER (Arthur).
Japan wax. [Abstract.] *Pharm. Journ.* III. x. (1880) 607–
608.
Beitraege zur Kenntniss pharmaceutisch wichtiger Gewaechse.
I. Ueber Smilax China L. und ueber die Sarsaparillwurzeln.
II. Ueber die Rhizome der officinellen Zingiberaceen, *etc.*
Halle, 1881. 8°.

MEYER (Ernst Heinrich Friedrich).
Ueber Seidenflachs, besonders den neuseelaendischen. Vorgelesen
in der physikalisch-oekonomischen Gesellschaft zu Koenigs-
burg, der 18 Febr. 1842 ; darauf mehrfach berichtigt und
erweitert. [Koenigsberg, 1842.] 8°.
Die Vertheilung der Nahrungspflanzen auf der Erde, *etc.*
[Koenigsburg, 1846.] 8°.

MEYER (Friedrich Albrecht Anton).

Dissertatio inauguralis de cortice Angusturæ. Goettingae, 1790. 8°.

MEYER (R.).

Chemische Verarbeitung der Pflanzen- und Thierfasern. Braunschweig, 1880. 8°.

MICHAUX (François André).

Histoire des arbres forestiers de l'Amérique septentrionale, considérees principalement sous les rapports de leurs usages dans les arts et de leur introduction dans le commerce, *etc.* Paris, 1810–13. 3 vols. 4°.

The North American Sylva, or a description of the forest trees of the United States, Canada, and Nova Scotia. Considered particularly with respect to their use in the Arts, *etc.* To which is added a description of the most useful of the European forest trees. [Transl. by A. L. Hillhouse.] Philadelphia, 1817–19. 3 vols. 8°. Ed. [3.] by J. J. Smith, 1850.

MIDDLETON (W. H.).

Manual of Coffee-planting. Durban, 1866. 8°.

MIERS (John).

On the history of the "Matè" plant, and the different species of Ilex employed in the preparation of the "Yerba de Maté," or Paraguay tea. *Ann. Nat. Hist.* viii. (1861) 219–228, 389–401.

Brazil nuts. [Extract from *Trans. Linn. Soc.*] *Pharm. Journ.* III. v. (1875) 726.

MILHAU (—.).

Dissertation sur le caffeyer. Montpellier, 1746. 8°.

Dissertation sur le cacaoyer. Montpellier, 1746. 8°.

MILLER (—.).

Deer tongue (Liatris odoratissima) in perfumery. *Pharm. Journ.* III. v. (1875) 793.

MILLER (A. W.).

Mezquite gum. [*Proc. Amer. Pharm. Assoc.*] *Pharm. Journ.* III. vi. (1876) 942–945.

MILLER (Joseph).

Botanicum officinale. London, 1722. 8°.

MIQUEL (Friedrich Anton Wilhelm).

De nord-nederlandsche vergiftige gewassen. Amsterdam, 1836. fol.

Leerboek tot de kennis der artsenijgewassen, krachten, gebruik, en pharmaceutischen bereidingen. Amsterdam, 1838. 8°.

(Bernelot)
kultuur in Azië . Batavia. 18

MIQUEL (Friedrich Anton Wilhelm) *continued :*—
>   Chinchonae in Java. [Extract.] *Pharm. Journ.* III. i.
>   (1870) 90.

MITCHELL (Charles L.).
>   The active principles of the official Veratrums. A chemico-
>   physiological study. Part I.—Botanical. *Pharm. Journ.* III.
>   v. (1875) 768–770.
>
>   Gum-hogg. [*Amer. Journ. Pharm.*] *Pharm. Journ.* III. xi.
>   (1880) 154–155.

MITCHELL (John).
>   Dendrologia, or a treatise on forest trees, with Evelyn's Sylva
>   revised, corrected and abridged. London, 1828. 8°.

MITSCHERLICH (Gustav Alfred).
>   De Cacao. D. Berolini, 1857. 8°.
>
>   Der Cacao und die Chocolade. Berlin, 1859. 8°.

MODLEN (R.).
>   Notes on Chian Turpentine. *Pharm. Journ.* III. x. (1880)
>   913–914.
>
>   Notes on the Aristolochiaceae as antidotes to snake poisons.
>   *Pharm. Journ.* III. xi. (1880) 411.

MOE (N.).
>   De norske Fodervaexter, *etc.* Christiania, 1852. 8°.

MOELLER (Josef).
>   Pflanzenrohstoffe. I. Gerb- und Farbmaterialien. II. Fasern.
>   Wien, 1879. 8°.

MOERMAN-LAUBUHR (Théophile).
>   La ramie ou ortie blanche sans dards, . . . plante textile . . .
>   sa description, son origine, *etc.* Gand, 1871. 8°.

MOHL (Hugo von).
>   Inquiries into the formation of Gum Tragacanth. [Transl.
>   from the *Bot. Zeitung*.] *Pharm. Journ.* xviii. (1859) 370–
>   375.

MOLITOR (Agóst).
>   Gazdag csentartalmú akáczfaz. [An Acacia rich in Tanning
>   power.] *Erdészeti lapok*, 1880, pp. 423–425.

MONARDES (Nicolás).
>   Historia medicinal de las casas que se traen de nuestra Indias
>   occidentales, que serven en medicina. Sevilla, 1569. 4°. Ed.
>   2. 1574. Ed. 3. (Prima y segunda y tercera partes de la
>   historia medicinal) 1580. 4°.
>
>   (For versions in Latin, English, Italian and French, see Pritzel's
>   Thesaurus, Ed. 2, pp. 221–222.)

MONEY (Edward).

A letter on the cultivation of Cotton . . . and other matters connected with India, *etc.* London, 1852. 8°.

The cultivation and manufacture of Tea. Calcutta, 1874. 8°. Ed. 3. London, 1878. Ed. [4.] 1881.

MONGREDIEN (Augustus).

Trees and shrubs for English plantations, *etc.* London, 1870. 8°.

The Heatherside Manual of hardy trees and shrubs; being an alphabetical catalogue of all the hardy trees and shrubs most worthy of cultivation; forming a dictionary of both the English and botanical names of ornamental and useful trees and shrubs, with occasional descriptive remarks, and the addition of the botanical synonyms. With an introduction, *etc.* London, 1874. 8°.

MONNEREAU (Élie).

Le parfait indigotier, ou description de l'indigo. Nouvelle édition, *etc.* Amsterdam, 1765. 12°.

MONTGOMERIE (William).

History of the Introduction of Gutta Percha into England. [From the *Mechanics' Magazine.*] *Pharm. Journ.* vi. (1847) 377–379.

MONZON (J. Rafael).

Notice of some vegetable and animal substances, natural products of New Grenada. [In a letter to Dr. Pereira.] *Pharm. Journ.* xi. (1851) 262–263.

MONZINI (Antonio).

La coltura del gelso. Milano, 1875. 16°.

MOODY (Sophy).

The Palm Tree. London, 1864. 12°.

MOORE (Charles).

On the woods of New South Wales. Sydney, 1871. 8°.

MOQUIN-TANDON ([Christian Horace Bénédict] Alfred).

Éléments de botanique médicale contenant la description des végétaux utiles à la medicine et des espèces nuisibles à l'homme, vénéneuses ou parasites précédée de considérations sur l'organisation et la classification des végétaux. Paris, 1861. 8°. Ed. 2. 1866.

MORDANT DE LAUNAY (Jean Claude Mien), & Jean Louis Auguste LOISELEUR-DESLONGCHAMPS.

Herbier général de l'amateur, contenant la description, l'histoire, les propriétés et la culture des végétaux utiles et agréables. Paris, 1816–28. 4°.

MOREAU DE JONNÈS (Alexandre).

Recherches sur les changemens produits dans l'ètat physique des contrées par la destruction des forêts. Bruxelles, 1825. 4°.

—— Untersuchungen ueber die Veraenderungen, die durch die Ausrotting der Waelder in dem physischen Zustande der Laender entstehen. Aus dem Franzoesischen von W. Widenmann. Tuebingen, 1828. 8°.

MOREL (Julius).

The turpentines and resinous products of the Coniferae. I. Strassburg turpentine. *Pharm. Journ.* III. viii. (1877) 21–22. II. Canada balsam, *Ib.* 22–23. III. Common turpentine, *Ib.* 81–83. IV. Venice turpentine, *Ib.* 281–283. V. Hungarian balsam, *Ib.* 283. VI. Carpathian balsam, *Ib.* 283. VII. Turpentine from the Aleppo pine, *Ib.* 283. VIII. Burgundy pitch, *Ib.* 342–344. XI. Galipot, *Ib.* 344. XII. Pitch, *Ib.* 344. XIII. Wood Tar, *Ib.* 344. XIV. Juniper Tar, *Ib.* 344 (542). XV. Oil of Turpentine, *Ib.* 543–545, 725–727. XVI. Oil of Juniper, *Ib.* 886–887. XVII. Oil of Savin, *Ib.* 981–982. XVIII. Oil of Cedar, *Ib.* 982. XIX. Colophony, *Ib.* 982-984. XX. Gum Sandarach, *Ib.* 1024–1025. XXI. Amber, *Ib.* (1879) 673–676. XXII. White Dammar, *Ib.* 714–716. XXIII. Kauri resin, *Ib.* 715–716.

MOREIRA (Nicolau Joaquim).

Diccionario de plantas medicinaes Brasilieras contendo o nome de planta, seu genero, especie, familia e o botanico que a classificacon ; o logar onde é mais commun, as virtudes que se lhe atribue e as doses e formas de sua applicação, *etc.* Rio de Janeiro, 1862. 8°.

Vocabulario das arvores Brasilieras que podem fornecer madeira para construcções civis, navaes e marcenaria, seguido de um indiculo botanico de algunas plantas de Paraguay. Rio de Janeiro, 1870. 8°.

Breves considerações sobre a historia e cultura do cafeiro e consumo de seus productos. Rio Janeiro, 1873. 8°.

U.S. Centennial Exhibition. Brazilian Coffee. New York, [1876]. 8°.

Historical notes concerning the vegetable fibres, exhibited by Severino L. da C. Leite. New York, 1876. 8°.

MOREL (L. F.).

Traité des champignons au point de vue botanique, alimentaire et toxologique. Moulins, 1865. 16°.

MORGAN (—.).

The Carnauba tree. [Extract from Consular Report.] *Pharm. Journ*. III. vi. (1876) 745–746.

MORREN (Charles François Antoine).

The Asparagus of the Cossacks (Typha latifolia). [From the *Journ. d'Hortic. Gand.*] *Pharm. Journ*. viii. (1848) 543–544.

MORRIS (D.).

Notes on Liberian coffee. Jamaica, 1881. fol.

MORRIS (Fred. W.).

The Sarracenia purpurea. A remedy for Small Pox. [From the *Amer. Med. Times*.] *Pharm. Journ*. II. iv. (1862) 87–88.

MORSON (T. R. N.).

Observations on certain plants of the genus Piper. *Pharm. Journ*. iii. (1843) 471–478, 525.

MOSELEY (Benjamin).

A treatise concerning the properties and effects of Coffee. London, 1785. 8°. Ed. 2. 1785. Ed. 3. 1785. Ed. 5. 1792.

MOSS (John).

Note on a sophistication of Pareira root. *Pharm. Journ*. III. iv. (1874) 911.

Structure and development of Pareira stem, Chondodendron tomentosum, R. et P. *Pharm. Journ*. III. vi. (1876) 702–707. *See also* pp. 716–717.

Curara, the proposed remedy for rabies. Synonyms: Ourari, Urari, Wourari, Wourali, Wouraly. *Pharm. Journ*. III. viii. (1877) 421–424.

MOTLEY (James).

On the use of coffee leaves in Sumatra. [Extract from a letter to Sir W. J. Hooker.] *Pharm. Journ*. xiv. (1855) 427.

MOURIER (Charles).

Anviisning til Traeplantning og Skovs Anlaeg paa en Bonde-gaardslod saavel til Garntoemmer som til Braendsel. Kjoeben-havn, 1806. 8°.

MUELLER (Anton)

Alphabetisches Woerterbuch synonymer lateinischer, deutscher und boehmischer Namen der offizinellen Pflanzen. Prag, 1866. 4°.

MUELLER (Ferdinand).

Das grosse illustrirte Kraeuterbuch. Ausfuehrliche Beschreibung aller Pflanzen, ihres Gebrauchs, Nutzens, ihre Anwendung und Wirkung in der Arzneikunde, ihres Anbaus, ihre Ein-sammlung, Verwerthung und Verwendung im Handel und Gewerbe, *etc.* Ed. 5. Ulm, 1877. 8°.

MUELLER (Ferdinand, *Baron* von).

Australian medicinal plants. [Extract from first report.] *Pharm. Journ.* xv. (1855) 114–116.

Australian · vegetation, indigenous or introduced, considered especially in its bearings on the occupation of the territory, with a view of unfolding its resources. Melbourne, 1867. 8°.

Intercolonial Exhibition Essays, 1866-67, No. 5.

Report on the vegetable products exhibited in the international exhibition of 1866–67. Melbourne, 1867. 8°.

On the application of Phytology to the industrial purposes of life. Melbourne, [1870]. 8°.

Forest Culture in its relation to industrial pursuits. Melbourne, [1871]. 8°.

The principal timber trees readily eligible for Victorian industrial culture, with indications of their native countries and technologic uses. (Report of the Acclimatisation Society of Victoria, pp. 29–58.) Melbourne, 1871. 8°.

The conservation of forests and the production of potash. [From *The Gardeners' Chronicle.*] *Pharm. Journ.* III. ii. (1872) 565–566.

The economic products of Eucalyptus trees. [Extract.] *Pharm. Journ.* III. ii. (1872) 628.

The objects of a botanic garden in relation to industries, *etc.* Melbourne, 1872. 8°.

Additions to the lists of the principal timber trees and other select plants, readily eligible for Victorian industrial culture. (Issued in 1871 and 1872, by the Acclimatisation Society.) Melbourne, [1872]. 8°.

Second supplement to the select plants, readily eligible for Victorian Industrial culture. Melbourne, [1874]. 8°.

Tea. A lecture delivered at Ballarat. [Ballarat], 1875. 12°.

Select plants readily eligible for industrial culture or naturalisation in Victoria; with indications of their native countries and some of their uses. Melbourne, 1876. 8°.

Eucalyptographia. A descriptive atlas of the Eucalypts of Australia and the adjoining islands. 1878, *etc.* 8°.—→

Suggestions on the maintenance, creation, and enrichment of forests. Melbourne, 1879. 8°.

Report on the forest resources of Western Australia. London, 1879. · 4°.

MUELLER (Ferdinand, *Baron* von) *continued :—*
Select extra-tropical plants, readily eligible for industrial culture
and naturalisation. Calcutta, 1880. 8°.

MUELLER (Hugo).
Die Pflanzenfaser und ihre Aufbereitung fuer die Tecknik.
Braunschweig, 1876. 8°.

MUELLER (Volkmer).
Deutsche Brennnesseln. Ihre Cultur und Verwerthung als
Gespinnst- und Handelspflanzen. [No. 160. Landwirthschaft-
liche Volksbuecher.] Leipzig, 1878. 8°.

MUENTER (Andreas Heinrich August Julius).
Ueber Tuscarora-rice (Hydropyrum palustre, L.). Griefswald,
1863. 8°.

MULLER [or MUELLER] (C. J.).
Preparations from Cannabis sativa in India. [Extract from a
letter.] *Pharm. Journ.* xiv. (1854) 165.

MUNIER (Jean Baptiste).
Quelques observations sur nos forêts de sapins. [From the
Journal du Jura.] Lons-le-Saulnier, 1867. 8°.

MUNRO (D. R.).
A description of the forest and ornamental trees of New Bruns-
wick. Saint John, N.B., 1862. 8°.

MURPHY (Edmund).
A treatise on the agricultural grasses. Dublin, 1844. 8°.

MURRAY (Andrew).
Mammoth trees [of California]. [Extract.] *Pharm. Journ·* II.
ii. (1861) 434–435.
The Pines and Firs of Japan. London, 1863. 8°.
The Food products of St. Petersburg. [Extract.] *Pharm. Journ.*
III. i. (1871) 788–789.

MURRAY (James P.).
Pituri. [*Lancet.*] *Pharm. Journ.* III. ix. (1879) 638.

MURRAY (John).
An account of the Phormium tenax: or New Zealand flax.
Printed on paper made from its leaves (bleached), with a
postcript (*sic*) on paper. Ed. 2. London, 1838. 8°.
A descriptive account of the Palo de Vaca, or Cow-Tree of the
Caracas. With a chemical analysis of the milk and bark.
London, 1837. 8°. Ed. 2. 1838.

NANQUETTE (Henri).
Cours d'aménagement des fôrets, professé à l'école forestière de
Nancy. Paris, 1860. 8°.

NARDO (Giovanni Domenico).
Considerazioni generali sulle alghe, loro charactere, classificazione compozitione chemica et applicazione alla medicina, all' arte, all' agricoltura, *etc.* Venezia, 1835. 4°.

NAUDIN (Charles).
Essai de culture du cotonnier précoce du Japon à la villa Thuret, d'Antibes. [Extr. *Bull. Soc. d'acclimatation.*] Paris, 1880. 8°.

NEANDER (Johann).
Tabacologia ; h.e. Tabaci seu Nicotianae descriptio medico-chirurgico-pharmaceutica, *etc.* Lugduni-Batavorum, 1622. 4°.

NEBEL (Wilhelm Bernhard).
De plantis dorsiferis usualibus. D. Heidelbergae, 1721. 4°.

NEES VON ESENBECK (Theodor Friedrich Ludwig).
Plantae officinales, oder Sammlung officineller Pflanzen, *etc.* Duesseldorf, 1821–33. fol.

NEES VON ESENBECK (Theodor Freidrich Ludwig), & Carl Heinrich EBERMAIER.
Handbuch der medizinisch-pharmaceutischen Botanik, *etc.* Duesseldorf, 1830–33. 3 vols. 8°.

NETTO (Ladislau de Sousa Mello).
Apontamentos sobre a colleçao das plantas economicas do Brasil para a exposiçao internacional de 1867. Paris, 1866. 8°.

NEUBRAND (J. G.).
Die Gerbrinde, mit besondrer Beziehung auf die Eichenschaelwald-Wirthschaft, fuer Forstwirthe, Waldbezitzer und Gerber, *etc.* Frankfurt-am-Main, 1869. 8°.

NEVINS (John Birkbeck).
On the introduction and cultivation of Chinchona in India. *Pharm. Journ·* II. iv. (1863) 549–558.

NEWBERY (J. Como).
Timbers of Victoria. A descriptive catalogue of the specimens in the Industrial and Technological Museum (Melbourne) illustrating the economic woods of Victoria. Melbourne, 1877. 8°.

NICHOLLS (N. A. A.).
The cultivation of Liberian coffee in the West Indies. London, 1881. 8°.

NICOL (Robert).
A treatise on coffee, its properties, and the best mode of keeping and preparing it. Ed. 2. London, 1831. 8°.

NICOL (Walter).

The practical planter ; or a treatise on forest planting, comprehending the culture and management of planted and natural
timber, etc.   Edinburgh, 1799.   8°.   Ed. 2. London, 1803.
8°.

NIEMANN (Johann Friedrich).

Pharmacopoea batava, cum notis et additamentis medico-pharma-
ceuticis.   Ed. 2.   Lipsiae, 1824.   2 vols.   8°.

NOBREGA (Gerardo Jose de).

On the cultivation of the cochineal.   [Directions for the cultivation of the Nopal, etc.]   *Pharm. Journ.* viii. (1849) 342–348.

NOEHDEN (Heinrich Adolph).

Entwurf zu Vorlesungen ueber die pharmacologische Botanik.
Goettingen, 1802.   8°.

NOERDLINGER (Hermann).

Die technischen Eigenschaften der Hoelzer, *etc.*   Stuttgart, 1860.
8°.

Querschnitte von hundert Holzarten umfassend die Wald- und
Gartenbaumarten, *etc.*   Stuttgart und Tuebingen, 1852.   8°.

Les bois employés dans l'industrie.   Descriptions accompagnées
de cent sections en lames minces des principales essences forestières de la France et de l'Algerie, *etc.*   Paris, 1872.   16°.

Deutsche Forstbotanik oder forstlich-botanische Beschreibung
aller deutschen Waldhoelzer sowie die haeufigeren oder interes-
santen Baeume und Straeucher unserer Gaerten und Parkan-
lagen.   Fuer Forstleute, Physiologen und Botaniker, *etc.*
Stuttgart, 1874, *etc.*   8°.

NOETHLICHS (J. L.).

Die Korbweiden-Kultur oder Anlage und Unterhaltung der
Korbweiden-Pflanzungen in den Neiderungen.   Weimar, 1875.
8°.

NOGUEIRA DA GAMA (Manoel Jacinto).

Memoria sobre o Loureiro Cinnamoms vulgo Canaleira de Ceylão.
Lisboa, 1797.   8°.

NOIROT BONNET (L.).

Considerations sur les fôrets, sur la nécessité et sur les moyens
d'augmenter la valeur de leurs produits.   Paris, 1819.   8°.

Traité de la culture des forêts, ou de l'application des sciences
agricoles et industrielles à l'economie forestière, *etc.*   Paris,
1832.   8°.   Ed. 2. 1839.

Théorie de l'aménagement des forêts, *etc.*   Ed. 2.   Paris, 1842.
8°.   Ed. 3. 1843.

NORMANN (J. E.).

Om Tobaksplantens Dyrkning i Norge, gjennemgaaet og foroeget af Mart. Richard Flor. Christiania, 1811. 8°.

NOTHNAGEL (Hermann), & Michael Joseph ROSSBACH.

Handbuch der Arzneimittellehre. Ed. 2. Berlin, 1874. 8°. Ed. 4. 1880.

NOURIJ (Franz Gustav).

Historia botanica, chemico-pharmaceutica et medica foliorum Diosmae serratifoliae (vulgo foliorum Buchu). D. Groningae, 1827. 8°.

NOURRIGAT (E.).

Le murier du Japon (Morus japonica). Paris, 1873. 8°.

NOUVEL (Auguste).

Le tabac. Traité des semis et des séchoirs, à l'usage des planteurs de la Dordogne. Brive, 1876. 8°.

NUTTALL (Thomas).

The North-American sylva, or a description of the forest-trees of the United States, Canada and Nova Scotia, not described in the works of François André Michaux, *etc.* Philadelphia, 1842–54. 3 vols. 8°.

OBERLIN (L.).

Aperçu systematique des végétaux medicinaux, des végétaux alimentaires, ainsi que des végétaux employés dans les arts et dans l'industrie, *etc.* Strasbourg, 1867. 8°.

OBERLIN (L.), & —. SCHLAGDENHAUFFEN.

Sur les écorces dites d'angusture vraie du commerce et principalement de l'angusture du Brésil. Nancy, 1877. 8°.

O'CONOR (J. E.).

Vanilla; its cultivation in India. Revised edition. Calcutta, 1875. 8°.

ODEPH (Alphonse).

Abrégé pratique sur la culture de l'opium indigène, pour servir de guide aux habitants des campagnes. [Extrait de *L'Union Pharmaceutique . . . centrale de France.*] Paris, 1862. 8°.

Traité complet de la culture de l'opium indigéne, *etc.* Luxeuil, 1865. 12°.

(There is also a German version by L. Graeter. Stuttgart, 1867.)

OELHAFEN VON SCHOELLENBACH (Carl Christoph).

Abbildung der wilden Baeume, Stauden- und Buschgewaechse, *etc.* Nuernberg, 1767–1804. 3 vols. 4°.

——— Traité des arbres, arbrisseaux et arbustes de nos fôrets, traduit de l'allemand par G. Benistant. Nuremberg, 1775. [3 parts only.] 8°.

OERSTED (Anders Sandoee).

Planterigets Naturhistorie, en almeenfattelig Fremstilling af de vigtigst Planter i deres Forhold til Menneskene og Jorden, *etc.* Kjoebenhavn, 1839. 8°.

Frilands Traevaexten i Danmark. Veiledning til Kundskab om de Traeer og Buske, som kunne dyrkes i Friland i Danmark. Kjoebenhavn, 1864–67. 8°.

Bidrag til Kundskab om Egefamilien i Nutid og Fortid. Kjoebenhavn, 1872. 4°.

The Silphium of the ancients. [*Journ. Bot.*] *Pharm. Journ.* III. iii. (1873) 1012–1013.

OGILVIE (J.).

Beetroot and Beetroot Sugar. Cork, 1877. 8°. *

OLCOTT (Henry S.).

Sorgho and Imphee, the Chinese and African sugar canes. A treatise upon their origin, varieties and culture, *etc.* New York, 1857. 8°.

OLIGSCHLAEGER (F. W.).

Calendarium pharmaceuticum, oder Anweisung zur richtigen Einsammlung der vegetabilischen Arzneistoffe. Elberfeld, [1830]. 4°.

OLIVER (Daniel), & John Reader JACKSON.

A Handbook to the Museum of Economic Botany at Kew. [London,] 1861. 8°. Ed. 5. 1871. Ed. 6. 1875.

> Ed. 1. is by D. Oliver alone.

OLIVER (S. P.).

The Papaw tree. [Letter in *Nature.*] *Pharm. Journ.* III. x. (1879) 68–69.

OLUFSEN (Oluf Chr.).

Anvisning til Hampens Dyrkning. Kjoebenhavn, 1809. 8°. Ed. 2. 1819.

Udtag af Anviisning til Hampens Dyrkning. Christiania, 1812. 8°.

Anweisung zum Hanfbau. Aus dem Daenischen uebersetzt von T. Friedlieb. Altona, 1812. 8°.

ONDAATJE (W. C.). Products of Ceylon. [Abstract.] *Pharm. Journ.* xiii. (1854) 425–426.

ORLI (P. H. F. B.). *See* BOURGOIN D'ORLI.

O'SHAUGHNESSY (W. B.).

Preparation of Indian Hemp, or Gunjah. Calcutta, 1839. 8°.

The Bengal Dispensatory, and companion to the Pharmacopoeia, *etc.* London, 1842. 8°.

OTTAVI (G. A.).
Monografia dei prati artificiale coltivati ed erba medica, trifoglio, lupinella et sulla. Ed. 3. Casale, 1879. 8°.

OTTOLANDER (K. J.).
Practisch handboek voor de ooftboomteelt in Nederland. Groningen, 1880. 8°.

OUDEMANS (Cornelius Anton Johan Abraham).
Handleiding tot de pharmacognosie van het planten- en dierenrijk. Ed. 2. Amst. 1880. 8°.

OURCHES (Charles d').
Aperçu général des fôrets. Paris, 1805. 2 vols. 8°.

PAGENSTECHER (Friedrich).
Ueber Linum catharticum Linn. D. Muenchen, 1845. 8°.

PAGLIA (Enrico).
Delle erbe nocive ed utile spontanee nei prati Mantovani con tavole analitiche e suggerimenti pratici sulla loro nomenclatura e coltivazione. Mantova, 1872. 8°.

PALMER (Edward).
Medicinal plants used by North American Indians. [*Amer. Journ. Pharm.*] *Pharm. Journ.* III. ix. (1879) 773–774.

PALMER (J. Dabney).
The Quinine-flower. [Sabbatia Elliottii, *Steud.* Extract.] *Pharm. Journ.* III. vii. (1876) 371–372.

PAPILLON-BARDIN (—.).
Du lin et de sa culture dans le département de Seine-et-Marne. Meaux, 1878. 8°.

PAPPE (Carl Wilhelm Ludwig).
A List of South African indigenous plants used as remedies by the colonists of the Cape of Good Hope. Cape Town, 1847. 8°. [*Anon.*]
Florae capensis medicae prodromus; or an enumeration of South African plants used as remedies by the colonists of the Cape of Good Hope. Cape Town, 1850. 8°. Ed. 2. 1859.
Buchu. [Diosma crenata, *DC.* From Pappe's Fl. med. capensis prod.] *Pharm. Journ.* x. (1851) 475.
Cape Gum. [Acacia horrida, *Willd.* Doornboom.] *Pharm. Journ.* x. (1851) 530.
Cape Aloes. [Aloe ferox, *Lam.*] *Pharm. Journ.* x. (1851) 521.
Silvae capensis, or a description of South African forest-trees and arborescent shrubs, used for technical and economical purposes. Cape Town, 1854. 8°. Ed. 2. London, 1862.

PAULLI (Simon).

A treatise on Tobacco, Tea, Coffee, and Chocolate, . . . translated by Dr. James. London, 1746. 8°.

PAULLINI (Christian Franz).

De Jalapa liber singularis, *etc.* Francofurti, 1700. 8°.

Μοσχοκαρυογραφία, seu nucis moschatae curiosa descriptio historico-physico-medica, *etc.* Francofurti et Lipsiae, 1704. 8°.

PAUQUY (Charles Louis Constant).

De la Belladone, considerée sous ses rapports botanique, chimique, pharmaceutique, *etc.* Thése. Paris, 1825. 4°.

PAYEN (Anselme).

Mémoire sur l'amidon . . . Sur les fécules des diverses plantes et leurs applications. Paris, 1839. 8°.

Notice of a microscopic vegetation which attacks crystallized sugar. [Glycyphila, *Mart.*] *Pharm. Journ.* xi. (1852) 311–313.

PAYNO (Manuel).

Memorie sobre el maguey mexicano y sus diversos productos. Mexico, 1864. 4°.

PEACHIE (John).

The compleat herbal of physical plants. London, 1694. 8°.

For other tracts by this author, see the 'Guide to the Literature of Botany,' p. 199.

PECK (Charles H.).

The Balsam Fir. [Extract.] *Pharm. Journ.* III. xi. (1880) 333–334.

PECKOLT (Theodor).

On the Culture of Bixa orellana and the preparation of Annatto. [Extract.] *Pharm. Journ.* II. i. (1859) 185–186.

Explicações sobre a collecção de pharmacognosia e chimica organica etc., enviada a Exposição nacional. Cantagallo, 1861. 8°.

Analyses de materia medica Brasileira. Rio de Janeiro, 1868. 8°.

Historia das plantas alimentares e de gozo do Brasil, contendo generalidades sobre a agricultura brasileira, a cultura, uso, e composição chimica de cada uma dellas. Rio de Janeiro, 1872, *etc.* 8°.—→

Notes on Brazilian drugs. Sicopira. (Bowdichia major, *Mart.*) *Pharm. Journ.* III. vii. (1876) 69–70.

Carica Papaya and Papayotin. [Extract.] *Pharm. Journ.* III. x. (1879) 343–346, 383–386.

Myroxylon peruiferum, *Linn. f. Pharm. Journ.* III. xi. (1881) 818–819.

PECORI (G. R.).

Considerazioni sulla coltura dell' ulivo. Firenze, 1880. 8°.

PEIXOTO (Domingos Ribeiro dos Guimaraens).

Dissertation sur les médicamens brésiliens que l'on peut substituer aux médicamens exotiques dans la pratique de la médicine au Brésil, etc. Thèse. Paris, 1830. 4°.

PÉLAGAUD (E.).

L'Eucalyptus, sa culture forestière et ses applications industrielles. Lyon, 1881. 8°.

PELLEGRINI (Vittorio).

Stato attuale dei boschi del distretto forestale di Caprino veronese provvedimenti relativi: memoria. Verona, 1875. 8°.

PELLETIER-SAUTELET (—.).

On the plant which yields Patchouly. [Pogostemon Patchouly.] Pharm. Journ. viii. (1849) 574–576.

PENISTON (William M.).

Paris universal exhibition of 1867. Catalogue of Contributions from the Colony of Natal. London, [1867]. 8°.

PENNEY (William).

Similarity in the medical properties of two species of Cotyledon. Pharm. Journ. xi. (1851) 66–67.

On a Lobelia used medicinally in Peru. Pharm. Journ· xiii. (1853) 14.

PEREBOOM (Nicolaus Ewoud).

Materia vegetabilis, systemati plantarum praesertim philosophiae botanicae inserviens. Lugduni Batavorum, 1787–88. 4°.

PEREIRA (Jonathan).

The elements of Materia medica (and Therapeutics). London, 1839–40. 8°. Ed. 2. 1842. 2 vols. Ed. 3. 1849–53. Ed. 4. [by A. S. Taylor and G. O. Rees] 1854–57. Ed. [5.] by Robert Bentley and Theophilus Redwood, 1872–4.

Manual of Materia medica . . . an abridgment of . . . Dr. P.'s Elements of Materia medica . . . by F. J. Farre, etc. London, 1865. 8°.

Introductory lecture on [modern discoveries in] Materia medica, delivered at the establishment of the Pharmaceutical Society, March 20. Pharm. Journ. i. (1842) 565–580.

On the varieties of Hyoscyamus. Pharm. Journ. ii. (1842) 122–124.

On the characters which respectively distinguish the fruits of Hemlock, Anise, and Fool's Parsley, with a notice of a case of poisoning by Hemlock fruit taken in mistake for Anise fruit. Pharm. Journ. ii. (1842) 337–341.

PEREIRA (Jonathan) *continued* :—

On the Ceylon Cardamom. *Pharm. Journ.* ii. (1842) 384–389.

On Grains of Paradise. *Pharm. Journ.* ii. (1843) 443–446.

Notice of a Chinese article of the Materia medica, called "Summer-plant-winter-worm." [Sphaeria sp.] *Pharm. Journ.* iii. (1843) 591–594.

Some observations on Potato-starch. *Pharm. Journ.* iii. (1843) 20–25.

Observations on the Chinese gall, called ' Woo-pei-tsze,' and on the gall of Bokhara, termed ' Gool-i-pista.' *Pharm. Journ.* iii. (1844) 384–387.

On Palm Sugar. *Pharm. Journ.* v. (1844) 64.

On the Moth on whose larva the New Zealand Sphaeria Robertsii grows. *Pharm. Journ.* iv. (1844) 204–206.

Notices of some rare kinds of Rhubarb which have recently appeared in English commerce. *Pharm. Journ.* iv. (1845) 445–450.

Further notice respecting Siberian and Bucharian Rhubarbs, with some remarks on Taschkent Rhubarb. *Pharm. Journ.* iv. (1845) 500–502.

On some vegetable and mineral productions of New Zealand. *Pharm. Journ.* v. (1845) 72–75.

Note on Banbury Rhubarb. *Pharm. Journ.* vi. (1846) 76–78.

On Alcornoque bark. [? Quercus Suber L.] *Pharm. Journ.* vi. (1847) 362–368.

On the fruit of Amomum Melegueta, *Roscoe. Pharm. Journ.* vi. (1847) 412–419.

On the Cardamoms of Abyssinia. *Pharm. Journ.* vi. (1847) 466–469.

On the colouring matters of Dutch or Cape Litmus. *Pharm. Journ.* ix. (1849) 12–14.

On the commercial varieties of Ginger. *Pharm. Journ.* ix. (1849) 212–214, 261–265.

On the commercial varieties of Turmeric. *Pharm. Journ.* ix. (1850) 309–313.

On the Amomum citratum, an undescribed species of large Cardamom. *Pharm. Journ.* ix. (1850) 313–314.

The Kosso, or Brayera anthelmintica. *Pharm. Journ.* x. (1850) 15–24.

On Myrospermum pubescens, and its medicinal products, Balsamito, Balsam of Peru, and White Balsam. *Pharm. Journ.* x. (1850) 230–234.

PEREIRA (Jonathan) *continued* :—

On the Myrospermum of Sonsonate, from which Balsam of Peru, White Balsam, and Balsamito are obtained. *Pharm. Journ.* x. (1850) 280–290.

On Kokum butter, or the concrete oil of Mangosteen. *Pharm. Journ.* x. (1851) 65–56.

Notice of Nag-Kassar (Mesua ferrea, *Linn.*). *Pharm. Journ.* x. (1851) 321.

Additional observations on Dutch cake Litmus. *Pharm. Journ.* x. (1851) 325–326.

On the flower-buds of Calysaccion longifolium. *Pharm. Journ.* x. (1851) 449–450. *See also* p. 321.

Additional information on the mode of preparing the Balsam of Sonsonate, commonly called the Balsam of Peru. *Pharm. Journ.* xi. (1851) 260–261.

On Mishmee-bitter, or Coptis Teeta, of Wallich. *Pharm. Journ.* xi. (1852) 294–296.

PERIN (Eugéne).

Culture du houblon. Moyen d'augmenter son rendement. Strasbourg, 1874. 8°.

—— Der Hopfenbau. Mittel den Ertrag zu vermehren. Strassburg, 1874. 8°.

PERIODICAL WORKS. *See* end of Alphabetical list of Authors.

PERNITZSCH (Heinrich).

Flora von Deutschlands Waeldern mit besondrer Ruecksicht auf praktische Forstwissenschaft. Leipzig, 1825. 8°.

PERONI (E.).

La coltivazione del gelso. Trattato pratico, *etc.* Brescia, 1863. 8°.

PERRON (R.).

Instructions et renseignments sur le graines et sur les plantes fourragéres les plus connues dans le commerce et dans l'agriculture. Paris, 1865. 32°.

PERROTTET (Georges Samuel).

Catalogue raisonné des plantes introduites dans les colonies françaises de Bourbon et de Cayenne et de celles rapportées vivantes des mers d'Asie et de la Guyane au jardin du roi à Paris. Paris, 1824. 8°.

Observations sur les essais de culture tentés au Sénégal et sur l'influence du climat par rapport à la végétation, *etc.* Paris, 1831. 8°.

Mémoire sur la culture des indigoféres tinctoriaux et sur la fabrication de l'indigo. Paris, 1832. 8°.

PERROTTET (Georges Samuel) *continued :—*

Art de l'indigotier ou traité des indigofères. tinctoriaux et de fabrication de l'indigo, suivi d'une notice sur le Wrightia tinctoria, et sur les moyens d'estraire de ses feuilles le principe colorant qu'elles contiennent. Paris, 1842. 8°.

Rapport . . . sur une mission dans l'Inde, à Bourbon, à Cayenne, à la Martinique et à la Guadeloupe, concernant l'industrie sérigéne et la culture du mûrier. Paris, 1842. 8°.

Lettre sur l'introduction du Vanillier à l'île de la Réunion. Pondichery, 1860. 8°.

PERSOON (Christian Hendrik).

Traité sur les champignons comestibles, contenant l'indication des espèces nuisibles, précédé d'une introduction à l'histoire des champignons. Paris, 1818. 8°.

—— Abhandlung ueber die essbaren Schwaemme, uebersetzt von Dierbach. Heidelberg, 1822. 8°.

PERTHIUS (— de).

Traité de l'aménagement et de la restauration des bois de la France. Paris, 1803. 8°.

PETERMANN (A.).

Recherches sur la culture de la betterave à sucre. Bruxelles, 1876. 8°.

PETZOLD (E.), & G. KIRCHNER.

Arboretum muscaviense. Ueber die Entstehung und Anlage des Arboretum Sr. K. Hoh. des Prinzen Friedrich der Neiderlande zu Muskau, nebst einem beschreibenden Verzeichniss der saemmtlichen in demselben cultivirten Holzarten, *etc.* Gotha, 1864. 8°.

PEYL (Joseph).

Die landwirthschaftliche Pilzkunde fuer Landwirthe, Forstmaenner, Gaertner und die Hausfrauen. Mit besonderer Beruecksichtigung der parasitischen Feinde und Zerstoerer der Oekonomie, Industrie-, Forst-, und Garten-Gewaechse, sowie der Nahrungssubstanzen. Prag, 1863. 8°.

PHILLIPS (Henry).

History of cultivated vegetables, *etc.* Ed. 2. London, 1822. 2 vols. 8°.

PHIPSON (T. L.).

On a colouring matter obtained from the Rhamnus Frangula (Black Alder). *Pharm. Journ.* xviii. (1858) 285–286.

Observations on the agricultural chemistry of the Sugar Cane. London, 1873. 8°.

PHOEBUS (Philipp).

Deutschlands Kryptogamische Giftgewaechse in Abbildungen und Beschreibungen. Berlin, 1838. 4°.

Die Delondre-Bouchardatschen China-Rinden. Giessen, 1864. 8°.

PICCALUGA (Giuseppe).

Sulla coltivazione del cotone. Brescia, 1863. 8°.

PICCONE (A.).

Primi studii per una monografia delle principali varietà di ulivo coltivato nella zona ligure, (provincie di Genova, Porto Maurizio e Massa-Carrera) pubblicati per cura del comizio agrario di Genova. Genova, 1880. 8°.

PIERPONT (J. de).

Traité des arbres et arbustes rustiques en Belgique. Bruxelles, 1865. 8°.

PIERRE (J. Isidore).

Recherches analytiques sur la valeur comparée des principales variétés de betteraves, *etc.* Caen, 1857. 8°.

Études théoriques et pratiques d'agronomie et de physique végétale. (Vol. ii. Plantes fourragéres. Graines et produits derivés; iii. Céréales; iv. Plantes industrielles. Recherches diverses.) Paris, 1869. 18°.

PIERRE (L.).

Flore forestière de la Cochinchine. Paris, [1880, *etc.*]. fol.—→

PIESSE (G. W. Septimus).

The art of Perfumery, and the methods of obtaining the odours of Plants, *etc.* London, 1855. 8°. Ed. 2. 1856. Ed. 3. 1862. Ed. 4. 1879.

—— Des odeurs, des parfumes et des cosmetiques . . . Édition française publiée avec le concours de l'auteur, par O. Reveil. Paris, 1865. 12°.

Lecture on Perfumes, Flower-farming, and the methods of obtaining the odours of plants, *etc.* London, 1865. 8°.

PINCKARD (W.).

Geum montanum, or Indian Chocolate root. *Pharm. Journ.* iv. (1844) 222.

PINCKERT (Friedrich August).

Anleitung zur Cultur und Benutzung der Hirse als Koerner-, und Futterpflanze. Leipzig, 1871. 8°.

Die neuesten, eintraeglichsten und den Boden am meisten bereichernden culturpflanzen im Betriebe der Landwirthschaft unserer Zeit, in ihrer Bedeutung, Cultur und Benutzung als Nahrungs-, Futter-, und Handelsgewaechse. Berlin, 1857–67. 8°.

Piso (William).

G. P. de medicina brasiliensi libri IV., *etc.* Lugduni Batavorum, 1648. fol. Ed. 2. 1658.

G. P. Historia medica Brasiliae. Novam editionem curavit et praefatus est Josephus Eques de Vering. Vindobonae, 1817. 8°.

Pissot (A.).

Les graines et les plantes d'essence forestières à l'exposition international 1878 à Paris. Paris, 1881. 8°.

Plá y Rave (E.).

Manual del cultivo de arboles forestieras. Madrid, 1880. 8°.

Planchon (Gustave).

Des Globulaires, au point de vue botanique et médical. Montpellier, 1859. 8°.

Des Quinquinas. Paris, 1864. 8°.

Les kermes du chêne, aux point de vue zoologique, commercial et pharmaceutique. Paris, 1864. 8°.

Les charactères des drogues simples et usuelles. Paris, 1868. 8°.

Striated Ipecacuanhas. [*Journ. de Pharm.*] *Pharm. Journ.* III. iii. (1873) 521–523, 642–643.

The study of Botany and Materia medica. [Abstract.] *Pharm. Journ.* III. iii. (1873) 743–745.

The teaching of vegetable Materia medica. [Abstract of lecture.] *Pharm. Journ.* III. iv. (1873) 401–403.

Hoang-nan bark. [*Journ. de Pharm.*] *Pharm. Journ.* III. viii. (1877) 364.

Studies of the genus Strychnos. [*Journ. de Pharm.*] *Pharm. Journ.* III. xi. (1880) 469–471, 491–492; (1881) 529–531, 589–592, 693–695, 754–755.

Plantes qui fournissent le Curare. [From the *Journ. de Pharm.*] Paris, 1880. 8°.

Sur les plantes, qui servent de base aux divers Curares. *Comptes rendus*, Paris, xc. (1880) p. 133.

Planchon (Jules Émile).

Des Hermodactes, au point de vue pharmaceutique. Thèse. Paris, 1856. 8°.

On Hermodactyls. [Abstract.] *Pharm. Journ.* xv. (1856) 465–468, 500–503.

La truffe, et les truffières artificielles. Paris, [1875]. 8°.

Planellas Giralt (José).

Ensayo de una flora fanerogámica Gallega ampliada con indicaciones acerca los usos médicos de las especies que se describen. Santiago de Compostela, 1852. 8°.

PLANK (—.).

Grundriss der Veterinaerbotanik. Muenchen, 1840. 8°.

PLANTA-REICHENAU (Adolph von).

Iva (Achillea moschata). [Extract.]　*Pharm. Journ.* III. i. (1871) 727.

PLATO (Carl Gottlieb).

Deutschlands Giftpflanzen, zum Gebrauche fuer Schuelen fasslich beschrieben. Leipzig, 1829–40. 8°.

PLATT (L.).

Éléments d'agriculture. Travaux agricoles, plantes oléagineuses, tinctoriales, textiles . . . Publié par Ad. Rion. Boulogne, 1877. 16°.

PLAUCHUD (E.).

Quelques mots sur l'olivier et sur l'huile d'olive. Forcalquier, 1879. 8°.

PLENCK (Joseph Jacob).

Bromatologia, seu doctrina de esculentis et potulentis. Viennae, 1784. 8°.

> Also a German edition, same place and date.

Toxicologia, seu doctrina de venenis et antidotis. Viennae, 1785. 8°.

> Also in German; and Italian, Venezia, 1789. 12°.

Icones plantarum medicinalium secundum systema Linnaei digestarum cum enumeratione virium et usus medici, chirurgici atque diaetaetici. Viennae, 1788–1812. 8 vols. fol.

POCKLINGTON (Henry).

The Microscope in pharmacy. *Pharm. Journ.* III. ii. (1872) 621–622, 661–662, 701–702, 782–783, 821–822, 841, 921, 1005–1006, 1025; *Id.* iii. 1–2, 42–43, 62–63, 81–83, 102–104, 161, 181, 281–283, 301–302, 341, 382–383; (1873) 542–543, 581–582, 663–664, 702–703, 761–762, 824–825, 990–992; iv. (1874) 549–550; v. (1874) 83–84, 221–222, 261–263, 301–303.

POEDERLE (Eugène Joseph Charles Gilain Hubert d'Obnen, *Baron de*).

Manuel de l'arboriste et du forestier belgiques; ouvrage extrait des meilleurs auteurs et soutenu d'observations faites dans différent pays, où l'auteur a voyagé. Bruxelles. 1772. 8°. Supplément, 1779. Ed. 2. 1788. Ed. 3. 1792. 2 vols.

POEHL (A.).

Ein Beitrag zur Quebrachofrage. St. Petersburg, 1880. 8°.

POGSON (—.).

Transmission of seeds [of Chenopodium Quinoa. From *The Gardeners' Chronicle*]. *Pharm. Journ.* II. ix. (1868) 342.

Pokorny (Aloys).

Plantae lignosae imperii austriaci. Oesterreichs Holzpflanzen. Eine auf genaue Beruecksichtung der Merkmale der Laubhoelzer gegruendete floristische Bearbeitung aller im oesterreichischen Kaiserstadte wildwachsenden oder haeufig cultivirten Baeume, Straeucher und Halbstraeucher. Wien, 1864. 4°.

Polak (J. E.).

Assafoetida. [Extract.] *Pharm. Journ.* II. viii. (1867) 607.

Pomet (Pierre).

Histoire générale des drogues simples et composées, traitant des plantes, des animaux et des mineraux. Paris, 1694. fol. Ed. [2.] 1735. 2 vols. 4°.

—— History of drugs. London, 1712. 4°.

—— Aufrichtiger Materialist und Spezereyhandler. Leipzig, 1717. fol.

Porcher (Francis Peyre).

The medicinal, poisonous, and dietetic properties of the crypto-gamic plants of the United States. New York, 1854. 8°.

Resources of the Southern fields and forests, medical, economical, and agricultural; being also a medical botany of the southern states; with practical information on the useful properties of the trees, plants, and shrubs. Charleston, 1863. Ed. [2.] 1869. 8°.

Porter (Andrew R.).

Sium latifolium, Gray. [*Amer. Journ. Pharm.*] *Pharm. Journ.* III. vii. (1876) 174. *See also* Nathan Rogers, pp. 433–434.

This writer probably intends Sium latifolium of Linnaeus.

Porter (George Richardson).

On the nature and properties of the Sugar Cane; with practical directions for the improvement of its culture and the manufacture of its products. London, 1830. 8°.

The Tropical Agriculturist. A practical treatise on the cultivation and management of various productions suited to a Tropical Climate. London, 1833. 8°.

The nature and properties of the Sugar Cane. . . . Second edition with an additional chapter on the manufacture of Sugar from the beet-root. London, 1843. 8°.

Tratado practico sobre el cultivo del Algodon. Traducido del Ingles por el doctor N. Milano. Caràcas, 1863. 8°.

Le cocotier. Paris, 1874. 8°. *

Posada-Arando (A.).

Vandellia diffusa. An emetic. [*Journ. de Pharm.*] *Pharm. Journ.* III. ii. (1872) 849.

Pouchet (Félix Archimede).

Essai sur l'histoire naturelle et médicale de la famille des Solanées. Thèse. Paris, 1827. 4°.

Histoire naturelle et médicale de la famille des Solanées. Rouen, 1829. 8°.

Traité élémentaire de botanique appliquée, contenant la description de toutes les familles végétales et celle des genres cultivés, ou offrant des plantes remarquables par leurs propriétés ou par leur histoire. Paris, 1835–36. 2 vols. 8°.

Prescott (Henry Paul).

Tobacco and its adulteration. London, 1858. 8°.

Strong drink and Tobacco smoke; the structure, growth and uses of malt hops, yeast, and tobacco. [Edited by T. H. Huxley.] London, 1869. 8°.

Prestele (Joseph).

Die wichtigsten Giftpflanzen Deutschlands in lebensgrossen Abbildungen. Friedberg, 1843. 8°. Atlas in fol.

Prestoe (H.).

Report on Coffee cultivation in Dominico. Trinidad, 1875. 8°. *

Proctor, *junior* (William).

Observations on the Sassy bark of Western Africa, and on the tree producing it. [From *Amer. Journ. Pharm.*] *Pharm. Journ.* xi. (1851) 271–276.

On Erythrophleum judiciale (the Sassy bark tree of Cape Palmas). [From *Amer. Journ. Pharm.*] *Pharm. Journ.* xvi. (1856) 233–237.

A succinct description of a collection of American drugs presented to the Museum of the Pharmaceutical Society. *Pharm. Journ.* xvi. (1856) 268–274.

Note on Caramania Gum. [*Amer. Journ. Pharm.*] *Pharm. Journ.* II. vi. (1865) 658.

Puhn (Johann Georg).

Materia venenaria regni vegetabilis. Lipsiae, 1785. 8°.

Pulteney (Richard).

Dissertatio inauguralis de Cinchona officinali Linnaei, sive cortice peruviano. Edinburgi, 1764. 8°.

Quekett (Edwin John).

On the relation existing between the structure of the organs of plants and certain vegetable products. *Pharm. Journ.* iii. (1844) 321–326.

Quihou (—.).

Rapport sur les principales cultures faites en 1869 au jardin d'acclimatation du bois de Boulogne. (Extrait du Bull. Soc. imp. d'acclimatation, Avril, 1870.) Paris, 1870. 8°.

Commission des cheptels. Instructions aux chepteliers. 5me section. Végétaux, plantes alimentaires. Paris, 1873. 8°.

Rapport sur les principales cultures faites en 1874 au jardin d'acclimatation du bois de Boulogne. Paris, 1875. 8°.

Rabuteau (A.), & A. Peyre.

M'Boundou or Icaja, an ordeal poison used at the Gaboon. [*Comptes rendus.*] *Pharm. Journ.* III. i. (1870) 187.

Rafinesque-Scamaltz (Constantino Samuel).

Medical Flora: or manual of the medical botany of the United States of North America. Philadelphia, 1828–30. 2 vols. 8°.

Rafn (Carl Gottlob).

Danmarks ach Holsteens Flora systematisk, physisk og oeconomisk, bearbeydet. Kjoebenhavn, 1796–1800. 2 vols. 8°.

Om Hoeravlens Vigtighed for Danmark, dens Tilstand paa Hoeravlings Institutet Lykkensstaede i Fyen, samt om Hoerrens bedre Dyrkning og Behandling i dette Institut fremfor andensteds i Danmark. Kjoebenhavn, 1803. 8°.

Vejledning for Bonden til at dyrke Hoer og at bearbeide den indtil Heglingen. Kjoebenhavn, 1806. 8°. Ed. 2. (by Hans Andreas Wueff,) 1830.

—— Anleitung fuer den Landmann zum Flachsbau und zur Behandlung des Flachses bis zum Hecheln. Kjoebenhavn, 1807. 8°.

—— Anwisning til linodling och linets handterande tills det haecklas, *etc.* Oerebro, 1811. 8°.

Raibaud-L'Ange (Henri).

Du tabac, culture et preparation en Provence. Paris, 1860. 8°.

L'olivier, sa culture et ses produits. Paris, 1861. 8°.

Ramel (P.).

L'Eucalyptus globulosus [*sic*] (Tasmanian blue gum-tree) gommier bleu de la Tasmanie. [*Bull Soc. imp. d'acclimatation.*] Paris, 1862. 8°.

L'Eucalyptus globulus (Tasmanian blue gum-tree) gommier bleu de la Tasmanie. [*Bull. Soc. imp. d'acclimatation.*] Paris, 1869. 8°.

RAMIREZ (Braulia Anton).

Diccionario de bibliografía agronómica y de toda clase de escritos relacionados con la agricultura; seguido de un indice de autores y traductores con algunos apuntes biográficos. Madrid, 1865. 8°.

RAMON DE LA SAGRA. *See* LA SAGRA.

RATZEBURG (Julius Theodor Christian).

Die Waldverderbniss oder dauernder Schade, welche durch Insektenfrass, Schaelen, Schlagen und Verbeissen an lebenden Waldbaeumen entsteht, *etc.* Berlin, 1865. 2 vols. 8°.

RAVERET-WATTEL (Casimir).

L'Eucalyptus. Rapport sur son introduction, sa culture, ses propriétés, usages, *etc.* Paris, 1872. 8°. Ed. 2. 1876.

The Genus Eucalyptus : its acclimatization and uses. [Extract.] *Pharm. Journ.* III. iii. (1872) 22–24, 43–45.

RAWSON (*Sir* R. W.).

Report upon the rainfall of Barbadoes and its influence on the Sugar crops 1847–71. Barbadoes, 1874. fol. *

READE (Oswald A.).

Cinchona cultivation in St. Helena. *Pharm. Journ.* III. iii. (1873) 903.

REALI (Agostino).

Gli alberi e gli arbusti del cirdario e dell' Appenino Camerte, memoria sulle loro utilità e sui loro pregi in rapporto all' industria, al commercio, alle arte ed al miglioramente del patrio suolo. Camerino, 1872. 8°.

REDWOOD (Theophilus).

Notices of specimens in the Museum of the Pharmaceutical Society. Aloes wood. Woorari, or Wourali poison. The Fruit of the Carob Tree ; or St. John's Bread. *Pharm. Journ.* iii. (1843) 74–79.

Gray's Supplement to the Pharmacopoeia, *etc.* London, 1847. 8°. Ed. 2. 1848. Ed. 3. (A supplement to the Pharmacopoeia.) London, 1857.

REED (Eugene L.).

Statice caroliniana. [*Amer. Journ. Pharm.*] *Pharm. Journ.* III. x. (1879) 304.

REED (William).

The History of sugar and sugar yielding plants, together with an epitome of every notable process of sugar extraction and manufacture, from the earliest times to the present. London, 1866. 8°.

REGGIO (E.).
> Catechismo per la coltivazione del cotone. Brescia, 1863. 8°.

REGIMBEAU (M.).
> Culture de la truffe. [Extract.] Nîmes, 1877. 8°.
>
> Le chêne yeuse ou chêne vert [Quercus Ilex] dans le Gard. Nîmes, 1879. 18°.

REGNAUD (Charles).
> Histoire naturelle hygiénique et économique du cocotier (Cocos nucifera, Linn.). Paris, 1856. 4°.

REGNAULT (François & Geneviève de Nangis).
> La botanique mise à la portée de tout le monde, ou collection des plantes d'usage dans la médicine, dans les alimens et dans les arts. [*Anon.*] Paris, 1774. 3 vols. fol.

REHMANN (Joseph).
> Beschreibung einer Thibetanischen Handapotheke. Ein Beitrag zur Kenntniss der Arzneykunde des Orients. St. Petersburg, 1811. 8°.

REICHEL (—.).
> On a commercial Cinchona bark, sold as yellow bark. [From the *Pharm. Centr. Blatt.*]. *Pharm. Journ.* viii. (1848) 192–193.

REICHENBACH (Anton Benedict).
> Die Pflanzen im Dienste der Menschheit. Monographien der wichtigsten Nutzpflanzen des In- und Auslandes in ihrer geschichtlichen botanischen, chemischen . . . Beziehung. Berlin, 1866. 8°.

REID (Hugo).
> Outlines of medical botany. Edinburgh, 1832. 8°. Ed. 2. 1839.

REIN (J. J.).
> Ginseng und Kampfer. Marburg, 1879. 8°.

REISSECK (Siegfried).
> Die Fasergewebe des Leines, des Hanfes, der Nessel und Baumwolle. Wien, 1852. fol.
>
> Die Palmen. Ein physiognomisch-culturhistorische Skizze. Wien, 1861. 8°.

REITLER (Johann Daniel von), & G. F. ABEL.
> Abbildung der hundert deutschen wilden Holzarten, *etc.* Stuttgart, 1790. 4°. Ed. [2.] 1805. Fortsetzung, 1803.

RENARD (Ed.).
> Note sur une nouvelle espéce de bambou, et sur des objets fabriqués avec ce végétal. Paris, [1875]. 8°.
>
> Les varechs, leur emploi en Europe et dans l'extrème Orient, dans l'industrie, en agriculture, *etc.* [Extrait.] Paris, 1878. 8°.

BENARD (Joseph Claudius).

Die inlaendischen Surrogate der Chinarinde, in besondrer Hinsicht auf das Kontinent von Europa. Mainz, 1809. 8°.

RENAULT (—.).

Des arbres résineux et de leur utilitè particulière pour le boisement des friches. Mirécourt, 1866. 12°.

RENOUARD (Alfred).

Le lin en Russie; culture, commerce, industrie. Lille, 1877. 8°.

Nouvelles recherches micrographiques sur le lin et sur le chanvre. Lille, 1878. 8°.

RENOUARD, *fils* (Alfred).

Études sur le travail des lins, chanvres, jutes, *etc.* Ed. 3. Lille, [1876]. 3 vols. 8°. Ed. [4.] Paris, 18 .

> "Ces 3 volumes sont publiés chacun avec une couverture et un titre spéciaux, lesquels ne font mention, ni du titre général de l'ouvrage, ni de la tomaison, ni de la circonstance que c'est une 4me edition, savoir:
>
> Le tome I sous le titre: Histoire de l'industrie linière en France, principalement à Lille et dans le departement du Nord.
>
> Le tome II sous le titre: Etudes sur la culture, le rouissage et le teillage du lin en France, Belgique, Italie, Hollande, Allemagne, etc.
>
> Le tome III sous le titre: Etudes sur le commerce du lin et la statistique linière en France, Angleterre, Belgique, Russie, Italie, Hollande, *etc.*"

De quelques essais rélatifs à la culture et à le préparation du lin. Lille, 1876. 8°.

Distinction du lin et du chanvre d'avec le jute et le phormium dans les fils et tissus, *etc.* Lille, 1875. 8°.

Statistique comparée de la culture du lin et du chanvre. (*Annales agronom.* vi. (1880) pp. 180–205.)

REVEL (—.).

Della necessità ed utilità di coltivare il cotone in Italia. Torino, 1863. 16°.

REUM (Johann Adam).

Grundriss der deutschen Forstbotanik. Dresden, 1814. 8°. Ed. 2. 1825. Ed. 3. 1837.

Die Deutschen Forstkraeuter. Ein Versuch, sie kennen, benutzen und vertilgen zu lernen. Dresden, 1819. 8°.

Oekonomische Botanik, oder Darstellung der haus- und landwirthschaftlichen Pflanzen, zum Unterrichte junger Landwirthe. Dresden, 1833. 8°.

Reuss (Christian Friedrich).

Kenntniss derjenigen Pflanzen, die Malern und Faerbern zum Nutzen gereichen konnen. Leipzig, 1776. 8°.

Reuter (Fr.)

Die Kultur der Eiche und der Weide in Verbindung mit Feldfruechten zur Erhoehung des Ertrages der Waelder, *etc.* Ed. 3. Berlin, 1875. 8°.

Rey (Ferdinand).

Note on Herba Santa Maria [Chenopodium sp.]. *Pharm. Journ.* III. ix. (1878) 713.

Reymond (M. L. A.).

Flore utile de la France, *etc.* Paris, 1854. 8°.

Reynard (—.).

La galéga, nouveau fourrage, sa culture. Montrottier, 1870. 16°.

Reynaud (J. M.).

La ramie, sa culture et son exploitation à l'île de la Réunion. Saint-Denis, 1881. 8°.

Reynaud (Joseph).

De l'olivier, sa culture, son fruit et son huile. Paris, 1862. 12°.

Reynoso (Alvaro).

Ensayo sobre el cultivo de la caña de azucar. Madrid, 1862. 8°. Ed. 2. 1865.

Apuntes acerca de varios cultivos cubanos. Madrid, 1867. 8°.

Riant (Aimé).

Le café, le chocolat, le thé. Paris, 1875. 32°.

Ribbentrop (Bernard).

Hints on Arboriculture in the Punjab, *etc.* Calcutta, 1874. 8°. (Also a Hindi version. [Lahore ?], 1874.)

Richard (Achille).

Histoire naturelle et médicale des différentes espèces d'Ipecacuanha en commerce. Paris, 1820. 4°.

Botanique médicale, ou histoire naturelle et médicale des médicamens, des poisons et des alimens, tirés du règne végétal. Paris, 1823. 2 vols. 8°.

—— A. R.'s Medizinische Botanik. Mit Zuzaetzen und Anmaerkungen von Gustav Kunze und G. F. Kummer. Berlin, 1824–26. 2 vols. 8°.

Élémens d'histoire naturelle médicale, contenant les notions générales sur l'histoire et les propriétés de tous les alimens, médicamens ou poisons, tirés des trois règnes de la nature. Paris, 1831. 2 vols. 8°.

RICHARD (Achille) *continued :*—
On Brazilian Sarsaparilla. [From the *Journ. de Chimie Méd.*] *Pharm. Journ.* iv. (1845) 427.

RICHARD (H.).
Die Gewinnung der Gespinnstfasern. Braunschweig, 1881. 8°.

RICHARDSON (D. L.).
Flowers and flower gardens. With an appendix of practical instructions and useful information respecting the Anglo-Indian flower garden. Calcutta, 1855. 8°.

RICHESCHI (Sigismondo).
Coltivazione della piante saccarine ed industria della zucchero in Italia; relazione al terzo congresse degli agricoltori italiani. Torino, 1873. 16°.

RIEHL (G. W. Fr.).
De Cassiae speciebus officinalibus. D. Halae, 1801. 8°.

RILAND (John).
Memoirs of a West India Planter. Published from an original MS. With a preface and additional details. London, 1827. 8°.

RIMMEL (Eugène).
The Book of Perfumes. London, 1865. 8°.

See Notice in *Pharm. Journ.* II. vii. (1865) 132-134.

A Lecture on the commercial use of Flowers and Plants, . . . London, [1865]. 8°.

RINIKER (Hans).
Ueber die Baumform und Bestandesmasse. Ein Beitrag zur forstlichen Statik. Aarau, 1872. 8°.

RISSO (Antoine).
Essai sur l'histoire naturelles des orangers, bigaradiers, limettiers, cedratiers, limoniers, ou citroniers cultivés dans le département des Alpes maritimes. Paris, 1813. 4°.

RISSO (Antoine), & A. POITEAU.
Histoire naturelle des Orangers. Paris, 1818-19. fol.

Historie et culture des Orangers. Nouvelle édition, entièrement revue et augmentée d'un chapitre nouveaux sur la culture dans le midi de l'Europe et en Algérie, par A. Du Breuil. Paris, 1872. fol.

RITTER (Carl).
Opiumcultur, und die Mohnpflanze. Berlin, [no date]. 8°.

Die Rhabarber (Rheum) nach ihrer Heimath, ihrem Handelsgang und der Sphaere ihrer Verbreitung in Hoch-Asien. Liebig, *Annal.* iii. (1832) 209-221.

RITTER (Carl) *continued :—*

Historisch-geographische und ethnographische Verbreitung der Theecultur, des Theeverkehrs, und Theeverbrauch, zumal auf dem Landwege, *etc.* Liebig, *Annal.* vi. (1833) pp. 88–108, 215-233.

Ueber die Asiatische Heimath und die Asiatische Verbreitungsphaere der Platane, des Olivenbaums, der Feigenbaums, der Granate, Pistacie, und Cypresse. Berlin, 1844. 8°.

Ueber die geographische Verbreitung des Zuckerrohrs, (Saccharum officinarum) und der alten Welt vor dessen Verpflanzung in die neue Welt. Berlin, 1840. 4°.

Ueber die Africanische Heimat der Caffeebaums (Coffea arabica). Berlin, *Bericht,* (1846) pp. 237–238.

Ueber die geographische Verbreitung von Kameel und Dattelpalme, *etc.* Berlin, *Bericht,* (1847) pp. 8–14.

Der Verbreitung der Dattelpalme in geographischer und ethnographischer Beziehung, so wie ueber ihre aelteste Cultur. Berlin, *Monatsber. Ges. Erdk.* vi. (1849) pp. 90–100.

Ueber die geographische Verbreitung der Baumwolle und ihr Verhaeltniss zur Industrie der Voelker alter und neuer Zeit. Berlin, [1851]. 4°.

RITTER (J. R.).

Die kaukasische Comfrey (Symphytum asperrimum). Eine neue Futterpflanze, die sich bewaehrt. Basel, 1880. 8°.

RIVETT CARNAC. *See* CARNAC.

RIVIÈRE (Auguste & Charles).

L'Eriodendron anfractuosum. Alger, 1875. 8°.

Le sésame. Alger, 1875. 8°.

Les Bambous, végétation, culture, multiplication en Europe, en Algérie, *etc.* Paris, 1879. 8°.

ROBBINS (Alonzo).

Maté, or Paraguay Tea. [*Amer. Journ. Pharm.*] *Pharm. Journ.* III. viii. (1878) 1027–1029.

ROBERTSON (B.).

Tea cultivation in Japan. [Extract from Consular report.] *Pharm. Journ.* III. vi. (1875) 128–129.

Japanese Lacquer. [Extract from Consular report.] *Pharm. Journ.* III. vi. (1875) 487–488.

Japanese vegetable and bees' wax. [Extract from Consular report.] *Pharm. Journ.* III. v. (1874) 584–585.

Indian *versus* Chinese Opium. [Extract.] *Pharm. Journ.* III. vii. (1876) 543–545.

ROBERTSON (James).

Remarks upon Atees [Aconitum heterophyllum, *Wall.*]. *Pharm. Journ.* xvii. (1858) 550.

ROBIN, *fils* (L.).

Mémoire sur le café, sur sa culture, son commerce, ses propriétés physiologiques, thérapeutiques et alimentaires. Abbeville, 1864. 18°.

ROBINSON (C. J.)

Our Salad herbs. *Pharm. Journ.* III. i. (1870) 167.

ROBINSON (S. H.).

The Bengal Sugar Planter. Being a treatise on the cultivation of the Sugar Cane and Date Tree in Bengal, and the manufacture of Sugar and Rum therefrom. Calcutta, 1849. 8°.

ROBINSON (William).

A descriptive account of Asam: with a sketch of the local geography, and a concise history of the Tea-plant of Asam, *etc.* Calcutta, 1841. 8°.

ROCHUSSEN (Jan Jacob).

Het wets-ontwerp op particuliere cultuur-ondernemingen in Nederlansche Indië, *etc.* 's Gravenhage, 1862. 8°.

Beantwoord op zijne brochure [*i.e.* the preceding]. Rotterdam, 1862. 8°.

Daniel Hooibrink's kuntstmatige bewerking in bevruchting van granen en boomen, en hare toepassing op cultures in Indië. 's Gravenhage, 1864. 8°.

—— Culture et fécondation artificielle des céréales et des arbres . . . traduit du hollandais par Émile Robin. Paris, 1864. 8°.

RODET (Henri J. A.).

Botanique agricole et médicale, ou étude des plantes qui intéressent principalement les medicins, les vétérinaires et des agriculteurs, *etc.* Paris, 1857. 8°. Ed. 2. 1872.

RODIE (H.).

Remarks on the Bebearu (or Greenheart) tree of British Guiana. [From the *Royal Gazette of Georgetown.*] *Pharm. Journ.* iv. (1844) 281–283.

RODICZKY (Eug. von).

Beitraege zur Geschichte, Statistik und Bibliographic der wichtigsten Culturpflanzen. Vol. i. Die Biographie des Kartoffel. Wien, 1878. 8°.

RODICZKY (Jenő).

Az igazi sáfrány műnelése. *Földmūnelesi Érdekenek,* 1880, pp. 52–53, 64–65.

RODIN (Hippolyte).

Les plantes médicinales et usuelles de nos champs, jardins, fôrets. Descriptions et usages des plantes comestibles, suspectes, vénéneuses, employées dans la médicine, dans l'industrie et dans l'économie domestique. Paris, 1872. 18°. Ed. 3. 1876.

ROEBBELEN (—.).

Die Bewaldung und sonstigen Melioration der Eifel im Reg. Bez. Trier, *etc.* Trier, 1876. fol.

ROESSIG (G.).

Convolvulaceae in medicinisch-pharmaceutischer Beziehung. Versuch einer pharmakognost. Monographie der Convolvulaceae. Leipzig, 1875. 8°.

ROESSLER-LADÉ (Auguste von).

Die Nessel eine Gespinnstpflanze. Mit Anleitung zur deren Anbau und weiteren Bearbeitung. Leipzig, 1878. 8°.

ROHDE (Michael).

Monographiae Cinchonae generis specimen, sistens historiam ejus criticam ad introductionem in hoc genus inserviente. D. Goettingae, 1804. 8°. Ed. 2. 1804.

ROHLFS (Gerhard).

Neue Beitraege zur Entdeckung und Erforschung Africa's. Cassel, 1881. 8°.

> Die Halfa und ihre wachsende Bedeutung fuer den Europaeischen Handel, pp. 20-30.

ROHR (Johann Andreas).

Resp. Dissertatio . . . de arundine saccharina. Vom Zuckerrohr . . . *Praes.* J. F. de Pre, *etc.* Erfordiae, [1719]. 4°.

ROHR (Julius Bernhard von).

Historia naturalis arborum et fruticum sylvestrium Germanicae, oder Naturgemaesse Geschichte der von sich selbst wilde wachsenden Baeume und Straeucher in Teutschland, *etc.* Leipzig, 1732. fol.

Physicalisch-oeconomischer Tractat von dem Nutzen der Gewaechse, insonderheit der Kraeuter und Blumen in Befoerderung der Glueckseligkeit und Bequemlichkeit des menschlichen Lebens, *etc.* Coburg, 1736. 8°.

Observations sur la culture du Coton, trad. de l'allemand. Paris, 1807. 8°.

ROHR (Julius Philipp Benjamin von).

Anmerkungen ueber den Cattunbau zum nutzen der Daenischen Westindischen Colonien. Altona und Leipzig, 1791-3. 2 pts. 8°.

Rojas Clemente (Simon de).
Memoria sobre el cultivo y cosecha del Algodon (Gossypium) en general y con aplicacion á España, particuliarmente á Motril. Madrid, 1818. 8°.

Romano (Antonio).
Plantae officinales in Europa sponte crescentés. Viennae, 1837. 8°.

Rondot (Natalis).
Notice du vert de Chine, et de la teinture en vert chez les Chinois. (Rhamnus utilis, *Decne.* Rhamnus chlorophorus, *Decne.*) Paris, 1858. 8°.

Roques (Joseph).
Plantes usuelles indigénes et exotiques, dessinées et coloriées d'après nature avec la description de leurs caractères distinctifs et de leur propriétés médicales. Paris, 1807-8. 2 vols. 4°.

Phytographie médicale . . . où l'expose l'histoire des poisons tirés du règne végétal, et les moyens de remédier à leurs effets délétères, avec des observations sur les propriétés et les usage des plantes heroïques. Paris, 1821. 2 vols. 4°. Ed. [2.] 1835. 3 vols. 8°.

Histoire des champignons, comestibles et vénéneux, où l'on expose leur charactères distinctifs, leurs propriétés alimentaires et économiques, leurs effets nuisibles et les moyens de s'en garantir ou. d'y remédier. Paris, 1832. 8°. Atlas, 4°. Ed. 2. 1841.

Nouveau traité des plantes usuelles, spécialement appliquées à la médicine domestique et au régime alimentaire de l'homme sain ou malade. Paris, 1837-38. 4 vols. 8°.

Rosenthal (David August).
Synopsis plantarum diaphoricarum, Systematische Uebersicht der Heil-, Nutz-, und Giftpflanzen aller Laender. Erlangen, 1862. 8°.

Rossignon (Jules).
Manual del cultivo del café, cacao, vainillo y tabaco en la America española, y de todas sus aplicaciones, *etc.* Paris, 1859. 12°. Ed. 3. Madrid, 1881. 18°.

Rostrup (Frederik Georg Emil).
Afbildning og Beskrivelse af de vigtigste Fodergraeser. En Vejledning for Landmaend til at laere Graesarterne at kjende og den hensigtsmaessigte Maade at benytte dem paa. Kjoebenhavn, 1865. 4°.

Rostrup (Frederik Georg Emil) *continued* :—
Sygdomme hos Skovtraeerne, fåraarsagede af ikke-rustagtige
Snystettesvampe. I. II. *Tidsskr. foer Skovbrug*, iv. (1879–80)
pp. 1–86, 113–206.

Ross (—.).
Persian Opium. [Extract from Consular report.] *Pharm. Journ.*
III. x. (1880) 1002, xi. (1881) 804.

Rossmaessler (E. A.).
Der Wald. Den Freunden und Pflegern des Waldes geschildert.
Leipzig, 1862. 8°. Ed. 2. (by M. Wilkomm), 1870. Ed. 3.
1880.

Roth (C. W.).
Handbuch zu der Bienenwirthschaftlichen Pflanzensammlung,
enth. 200 Pflanzen von denen die Bienen vorzugsweise ihre
noethigen Stoffe entnehmen, *etc.* Goettingen, 1874. 8°.

Herbarium to above in folio.

Roth (—.).
On oil of sesamum (Sesamum orientale), and its employment
in Pharmacy. [Extract.] *Pharm. Journ.* II. iii. (1861)
164.

Roth (Henry Ling).
A report on the sugar industry of Queensland. Brisbane,
1880. 8°.

Rothrock (Joseph T.).
Notes on economic botany of the Western United States.
[Extract.] *Pharm. Journ.* III. x. (1880) 664–666.

Rottboell (Christen Friis).
Descriptiones rariorum plantarum (surinamensium), nec non
materiae medicae atque oeconomicae e terra surinamensi frag-
mentum. D. Havinae, 1776. 4°. Ed. 2. (title varied).
Hafniae et Lipsiae, 1798. fol.

Rottenburg (V. H.).
Flore médicale belge. Bruxelles, 1881. 8°.

Roughley (Thomas).
The Jamaica Planter's Guide ; or a system of planting and managing a Sugar Estate, or other plantations in that Island, and
throughout the British West Indies in general, *etc.* London,
1823. 8°.

Rousse (Hippolyte).
Cultures du cotonnier, du lin et du chanvre. Blidah, 1864.
8°.

Roussel (Ernest).

Des champignons comestibles et vénéneux, qui croissent dans les environs de Paris. Paris, 1860. 8°.

Roussel (Henri François Anne de).

Tableau des plantes usuelles rangées par ordre, suivant les rapports de leur principes et de leurs propriétés. Caen, 1792. 8°. Ed. 2. [1797].

Rousset (Antonin).

Culture, exploitation et aménagement du chêne-liège en France et en Algérie, suivi d'un état detaillé des fôrets de chêne-liège de l'Algérie. Paris, 1859. 8°.

Recherches expérimentales sur les écorces à tan du chêne yeuse, relativement à la production et à l'aménagement des fôrets de cette essence. Paris, 1878. 4°.

Routledge (Thomas).

Bamboo and its treatment, cultivation, and cropping. Sunderland, 1879. 8°.

Bamboo, considered as a paper-making material, with remarks upon its cultivation and treatment. Supplemented by a consideration of the present position of the Paper trade in relation to the supply of the raw material. London, 1875. 8°.

Roxburgh (William).

A botanical description of a new species of Swietenia, with experiments and observations on the bark thereof, in order to determine and compare its powers with those of Peruvian bark, for which it is proposed as a substitute. [London, 1793.] 4°.

Royle (John Forbes).

An essay on the antiquity of Hindoo medicine, including an introductory lecture to the course of Materia medica and Therapeutics, delivered at King's College. London, 1837. 8°.

Essay on the Productive Resources of India. London, 1840. 8°.

On East Indian Kino. *Pharm. Journ.* v. (1846) 495–500.

Abstract of a paper on the Kino of Butea frondosa. *Pharm. Journ.* v. (1846) 500–501.

On the tree yielding African Olibanum. *Pharm. Journ.* v. (1846) 541–547.

A manual of Materia medica and Therapeutics, . . . with many new medicines. London, 1847. 12°.

Report on the culture of the China Tea Plant in the Himalayas. London, 1849. fol.

Royle (John Forbes) *continued :—*

On the culture and commerce of Cotton in India, and elsewhere. London, 1851. 8°.

On Indian Fibres fit for Textile Fabrics, or for Rope and Paper-making. *Journ. Soc. Arts,* ii. (1854) 366–372.

Materials for paper-making procurable from India. [Extract from *Journ. Soc. Arts.*] *Pharm. Journ.* xiv. (1854) 168–170.

The fibrous plants of India, fitted for cordage, clothing and paper. With an account of the cultivation and preparation of flax, hemp, and their substitutes. London, 1855. 8°.

Indian Fibres, being a sequel to Observations "On Indian Fibres fit for Textile Fabrics, or for Rope and Paper-making." *Journ. Soc. Arts,* v. (1856) 17–28.

Review of the measures which have been adopted in India for the improved culture in Cotton. London, 1857. 8°.

On cotton culture in India. [Notice.] *Pharm. Journ.* xvii. (1858) 494–496.

Roze (L.).

La menthe poivrée, sa culture en France, ses produits, falsifications de l'essence, et moyens de la reconnaître. Paris, 1868. 12°.

Rueckert (Ernst Ferdinand).

Beschreibung der am haeufigsten wildwachsenden und kultivirten phanerogamen Gewaechse, Farnkraeuter, so wie einiger offizinellen Moose und Schwaemme Sachsens und der angraenzenden Preussischen Provinzen, mit Angabe ihrer nuetzlichen und schaedlichen Eigenschaften. Fuer Freunde der Botanik, Schullehrer und Oekonomen bearbeitet. Leipzig, 1840. 2 vols. 8°.

Rufin (Alfred).

Der Flachsbau des Erdballs. Studien ueber die Naturgeschichte, Geographic, Geschichte, und Statistik des Flachses und der Flachs-Cultur aller Cultur-Laender der Erde. Ein Handbuch fuer Landwirthe und Industrielle, fuer landwirthschaftliche Vereine und Lehranstalten, sowie fuer die gesammte National-oeconomic. Berlin, 1878. 2 vols. 8°.

Ruiz (Sebastian Joseph Lopez).

Defensa y demonstracion del verdadero descubridor de las quinas del reyno de Sante Fé, con varias noticias útiles de este específico, en contestacion á la memoria de Don Francisco Antonio Zea. Madrid, 1802. 4°.

Ruiz Lopez (Hipolito).

Quinologia, o tratado del àrbol de la quina o cascarilla, con su descripcion, y la de obras especies de quinos nuevamente descubiertas en el Perú ; del modo de beneficiarla, de su eleccion, comercio, virtudes, y extracto elaborado con cortezas recientes, y de la eficacia de este, comprobada con observaciones ; á qui se añaden algunos experimentos chimicos, y noticias acerca del análisis de todas ellas. Madrid, 1792. 4°.

—— Von dem officinellen Fieberrindenbaum, und den andern Arten desselben . . . Zuerst aus dem Spanischen im Italienische, und aus diesem im Deutsche uebersetzt. Goettingen, 1794. 8°.

Supplementa á la Quinologia. Madrid, 1801. 4°.

Disertacion sobre la raiz de la ratánhia, específico singular contra los fluxos de sangre, *etc.* Madrid, 1799. 4°.

Memoria sobre las virtudes y usos de la planta llamada en el Perú Bejuco de la Estrella [Aristolochia fragrantissima]. Madrid, 1805. 4°.

Memoria de las virtudes y usos de la raiz de la planta llamada Yallhoy en el Perú [Monnina polystachya]. Madrid, 1805. 4°.

Memoria sobre la legítima Calaguala y obras dos raices que con el mismo nombre nos vienen de la América meridional [Polypodium Calaguala, *Fée*]. Madrid, 1805. 4°.

Russell (Logan Dillon Hooper, & Robert).

Jamaica: a home for the invalid, and a profitable field for the industrious settler. *Proc. Royal Colonial Institute,* x. (1879) 209–241.

Rutter (Henry).

Silk and Tea tables, showing their cost per pound as purchased in China and Japan, and as laid down in England, *etc.* London, 1868. 2 pts. 8°.

Ryan (John).

Camelina sativa, or common Gold of Pleasure. *Pharm. Journ.* x. (1850) 145–157.

Ryder (Thomas).

Some account of the Maranta, or Indian Arrow root, in which it is considered and recommended as a substitute for starch prepared from corn. London, 1796. 8°.

S., (J. L.).

Sugar, how it grows and how it is made, *etc.* London, [1844?] 4°.

SABBATINI (Rinaldo).

Propagazione della pianta da foraggio, Edisaro pratense per formare praterie artificiali, modo di coltivarla e farne la sollecita a propagazione. Udine, 1878. 16°. [Ed. 2.?] Poggibonsi, 1880. 8°.

Privately printed.

SABLON (—. Comte de).

Du reboisement des montagnes et de la culture forestière dans le département du Rhône. Lyon, 1880. 8°.

SABONADIÈRE (William).

The Coffee Planter in Ceylon. London, 1866. 8°. Ed. 2. 1870.

SACCHERO (Giacomo).

Utilità dell' Eucaliptus. Catania, 1868. 8°.

SAFFRAY (—.).

Les remèdes des champs; herborisations pratiques à l'usage des instituteurs . . . et de tous ceux qui donnent leurs soins aux malades. 1re partie, octobre à mars. 2e partie, avril à septembre. Paris, 1875. 2 vols. 32°. Ed. 2. 1876.

SAGOT (Paul).

De l'état sauvage et des résultats de la culture et de la domestication. Nantes, 1865. 8°.

Exploitation des fôrets de la Guyane française. Paris, 1869. 8°. *

Du manioc. Bull. Soc. bot. de France, xviii. (1871) 341–354.

L'arbre à pain. Paris, 1872. 8°.

Légumes et cultures potagères de la Guyane française. (Journ. Soc. centrale d'hort. de France, 2e serie, 6, 1872.) *

Le bananier. Paris, 1872. 8°.

Agriculture de la Guyane française. Paris, 1873. 8°. *

Des plantes oléagineuses cultivées à la Guyane française. Paris, 1873. 8°.

Remarques générales sur les plantes alimentaires à la Guyane. Paris, 1873. 8°.

The Manioc, or Tapioca plant. [Abstract.] Pharm. Journ. III. iii. (1873) 569.

SAGRA (Ramon de la). See LA SAGRA.

ST. GEORGE (—.).

On a very inferior Cinchona bark found in Commerce. [From the Jahrb. fuer Prakt. Pharm.] Pharm. Journ. v. (1846) 370.

SAINT-MOULIN (V. J. de).

Commentatio botanico-oeconomica de quibusdam arboribus in Belgio cultis. Trajecti ad Rhenum, 1827. 8°.

SAINT-YVES (A.).

De l'utilité des algues marines. Paris, 1879. 8 .

SALDANHA DA GAMA, *Filho* (José de).

Configuração e descripção de todos os orgãos fundamentaes das principaes madeiras de cerne et brancas da provincia do Rio de Janeiro e suas applicacões não engenharia, industria, medicina e artes, *etc.* Rio de Janeiro, 1864–72. 3 ptie. 8°.

Desenhos dos vegetaes que acham-se descriptos no primeiro volume ·da obra configuração e descripção de todos os orgãos fundamentaes das principaes Madeiras de cerne et branca da provincia do Rio de Janeiro. Rio de Janeiro, 1865. 4°.

Travaux au sujet des produits qui sont à l'exposition universelle de Paris en 1867. Paris, 1867. 8°.

Classement botanique des plantes alimentaires du Brésil. Paris, 1867. 4°.

Quelques mots sur les bois du Brésil qui doivent figurer à l'exposition universelle de 1867. Paris, 1867. 8°.

Catalogue of the products of the Brazilian Forests, at the International Exhibition in Philadelphia. New York, 1876. 8°.

Notes in regard to some textile plants of Brazil, at the International Exhibition in Philadelphia in 1876. New York, 1876. 8°.

SALOMON (—.).

Traité de l'aménagement des fôrets. Mulhouse, 1837–38. 2 vols. 8°.

SAMANOS (Eloi).

Traité de la culture du pin maritime, comprenant des études sur la création des forêts, leur entretien, leur exploitation, et la distillation des produits résineux. Paris, 1863. 8°.

SANGIORGIO (Paolo).

Istoria delle piante medicali. Milano, 1809–10. 4 vols. 8°.

SANTA MARIA (Fernando de).

Manual de medicinas caseras para consuelo de los pobres Indios en las provincias, y pueblos donde no ay médicos in botanico. [St. Thomas da Manila], 1815. 8°.

SARGENT (Charles S.).

Les fôrets du Nevada central, avec quelques remarques sur celles des régions adjacentes. (*Ann. Sci. Nat.*, sér. 6, vol. ix. pp. 32–46.)

SATTLER (Christopher Wilhelm).

Exercitatio . . . de infusi Veronicae efficacia preferendae herbae Thée, *etc.* *Praes.* F. Hoffmann. Halae Magdeburgicae, [1694]. 4°.

SAUNDERS (R.).

Cultivation of Opium in India. [Extract.] *Pharm. Journ.* III. iv. (1874) 652–654.

SAUNDERS (William).

Some medicinal plants of Canadian growth. [*Proc. Amer. Pharm. Assoc.*] *Pharm. Journ.* III. ii. (1871) 145–146.

Insect Powder. [Pyrethrum. Extract.] *Pharm. Journ.* III. x. (1879) 23–24.

SAWER (John Charles).

Notes on Patchouli. *Pharm. Journ.* III. xi. (1880) 409–411.

Opinions as to the origin of commercial Vanilla. *Pharm. Journ.* III. xi. (1881) 773–775.

SAWYER (James).

Notes on Gelsemium sempervirens. *Pharm. Journ.* III. iv. (1874) 998–999.

SCALING (William).

The Salix or Willow . . . Practical instructions for planting and culture; with observations on its value and adaptability for the formation of hedges and game coverts. Ed. 2. London, 1871. 8°.

SCHAEFFER (Jacob Christian).

Versuche und Muster ohne alle Lumpen oder durch mit einem geringen Zusatze derselben Papier zu machen. Regensburg, 1765. 2 vols. 4°.

Illustrated by nineteen samples.

SCHAEFFER (Martin Gottlob).

Abhandling om Skoves Opelskning med Hensyn til den danske Flaades Skibsbyggeri. Kjoebenhavn, 1811. 8°.

SCHAFRANOW (N. S.)

The growth of forests. [In Russ.] St. Petersburg, 1875. 8°.

SCHEFFMAYER (K.).

Tobacco and its culture. (Annual Report of Agricultural Department.) Madras, 1879. 8°. *

SCHENK (—.).

On Ervalenta. [From Buchner's *Archiv.*] *Pharm. Journ.* x. (1850) 309–310.

SCHERZER (Karl von).

Smyrna. Mit besonderer Ruecksicht auf die geographischen, wirthschaftlichen, und intellectuellen Verhaeltnisse von Vorder-Kleinasien, *etc.* Wien, 1873. 8°.

Vegetabilische Erzeugnisse und Handel damit, pp. 114-156.

SCHERZER (Karl von) *continued :—*

—— Smyrne, considerée au point de vue géographique, économique et intellectuel . . . Traduit de l'allemand par F. Silas,
etc. Vienne, 1873. 8°. Ed. 2. Leipzig, [1880].

> Produits du règne végétal, ed. 2, pp. 104–144.

Commercial notes on Opium. [Extract.] *Pharm. Journ.* III. xi.
(1881) 835–836.

SCHLAGINTWEIT (H. von).

A few species of Rhubarb. [Abstract.] *Pharm. Journ.* III. xi.
(1880) 454.

SCHINDLER (Carl).

Schematismus und Statistik der Staatsforste . . . des oesterreichischen Kaiserthums. Wien, 1864. 8°.

SCHLEIDEN (Matthias Jakob).

On the different kinds of fecula. [From the *Pharm. Central
Blatt.*] *Pharm. Journ.* iv. (1844) 89–91.

Handbuch der medizinisch-pharmaceutischen Botanik und botanischen Pharmacognosie. Leipzig, 1852–57. 2 vols. 8°.

Fuer Baum und Wald. Eine Schutzschrift an Fachmaenner und
Laien gerichtet. Leipzig, 1870. 8°.

SCHLICHTKRULL (Olof Nicolai Christopher).

De officinelle Planter ordnede efter De Candolle's naturlige Plante-
system, telligemed de vigtigste Characterer paa disse Planters
Familier. Kjoebenhavn, 1831. 8°.

SCHMIDLIN (Eduard).

Abbildung und Beschreibung der wichtigsten Futtergraeser und
Wiesenkraeuter nebst Angabe ihrer Cultur und ihres Nutzens,
etc. Esslingen, 1850. 8°. Ed. 2. 1868. Ed. 3. (by Wilh.
Schule, sen. & jun.) Stuttgart, 1877.

SCHMIDT (David).

Taschenbuch der pharmaceutisch-vegetabilischen Rohwaarenkunde
fuer Aertze, Apotheker und Droguisten. Jena, 1847. 8°.

SCHMIDT (Franz).

Oesterreich's allgemeine Baumzucht, oder Abbildungen in- únd
auslaendischer Baeume und Straeuche, deren Anpflanzung in
Oesterreich moeglich und nuetzlich ist. Wien, 1792–1822.
4 vols. fol.

SCHMIDT (Johann Karl).

Allgemeine oekonomisch-technische Flora, oder Abbildungen und
Beschreibungen aller in Bezug auf Oekonomie und Technologie
merkwuerdigen Gewaechse. Jena, 1827. 2 vols. 8°.

SCHMITT (Adolf).

Anlage und Pflege der Fichten-Pflanzschulen. Weinheim, 1875. 8°.

SCHOBER (J. H.).

De teelt van hop op heidegrond. Haarlem, 1865. 8°.

SCHOMBURGK (Richard).

The grasses and fodder plants which may be beneficial to the squatter and agriculturist in South Australia. Adelaide, [1873]. 8°.

Papers read before the Philosophical Society and the Chamber of Manufactures. Adelaide, 1873. 8°.

> Chiefly on Forestry, Tobacco, Silk, Sunflower, Urari.

The Guaco plant. [Extract from Report.] *Pharm. Journ.* III. iii. (1873) 1019.

Flower farming in South Australia. [Extract.] *Pharm. Journ.* III. x. (1879) 8.

SCHOMBURGK (Robert Hermann).

Cedrone seed. [From *The Athenaeum.*] *Pharm. Journ.* x. (1851) 518–519.

The Urari, or arrow poison of the Indians of Guiana. Remarks in continuation to those contained in the seventh volume of the Annals of Natural History. *Pharm. Journ.* xvi. (1857) 500–507.

The Vegetable products of Siam. [From *The Technologist.*] *Pharm. Journ.* II. iii. (1861) 123–128.

SCHRADER (Heinrich Adolph).

Grundriss meiner Vorlesung ueber die oekonomische Botanik. Goettingen, 1798. 8°.

SCHRADER (Johann Christian Karl).

Die norddeutschen Arzneipflanzen fuer Anfaenger der Apotheker-kunst. Berlin, 1792. 8°.

SCHREBER (Daniel Gottlieb).

Vom perennirenden sibirischen Leine und dessen auch bey uns mit Nutzen einzufuehrenden Baue handelt vorgaenzig, *etc.* [Halle], 1754. 4°.

SCHROEDER (Julius).

Das Holz der Coniferen. Dresden, 1872. 8°.

SCHROFF (Karl D.).

Notes on the Chilian drugs of the Novara expedition. [Trans. and abridged by Daniel Hanbury.] *Pharm Journ.* II. iv. (1863) 366–371.

SCHROTTKY (Eugène C.).

The Principles of rational Agriculture applied to India, and its staple products. Bombay, 1876. 8°.

SCHUBARTH (Heinrich).

Anweisung zum Anbau der bekanntesten, in Deutschland akklimatisirten Handelgewaechse, welche sich vorzuegl. zum Anbau auf der Felde in Grossen eignen, und zu deren Bereitung als Kaufmannswaare. Leipzig, 1825. 8°.

SCHUCHARDT (Theodor).

On Rhatany Root. [Transl. from the *Botanische Zeitung.*] *Pharm. Journ.* xvi. (1856) 29–32, 132–124.

SCHUEBELER (F. Christian).

Die Kulturpflanzen Norwegens, *etc.* Christiania, 1862. 4°.
—— Synopsis of the vegetable products of Norway. Transl. . . . by M. R. Barnard. 1862. 4°.

SCHULTZ (Franz Johann).

Abbildung der in- und auslaendischen Baeume, Standen und Straeuche, welche in Oestreich fortkommen, *etc.* Wien, 1792–1804. 3 vols. (and part of a fourth). fol.

SCHULZ (Franz).

Deutschlands Nutz- und Zierpflanzen. Deutschlands Waelder und Haine. Naturgeschichte der heimischen und harten auslaendischen Holzgewaechse, *etc.* Berlin, 1868. 8°.

SCHULZ (Johann).

De radice Ginseng vel Ninsi (Panax quinquefolium). D. Dorpati, 1836. 8°.

SCHULZE (A. G. R.).

Compendium der officinellen Gewaechse nach naturlichen Familien geordnet. Berlin, 1840. 8°.

SCHULZE (Richard).

Die Kultur der Korbweide. Brandenburg, 1874. 8°.

SCHUMACHER (Christian Friedrich).

Medicinisk Plantelaere for studerende Laeger og Pharmaceuter. Kjoebenhavn, 1825–26. 2 vols. 8°.

SCHUMACHER (Christian Friedrich), & Johan Daniel HERHOLDT.

De officinelle Laegemidler af Planteriget, som voxe vildt eller kunne dyrkes i de danske Stater. Kjoebenhavn, 1808. 4°.

SCHUMMEL (Theodor Emil).

Ueber die giftige Pilze, mit besondrer Ruecksicht auf Schlesien. Breslau, 1840. 4°.

SCHWALBE (Christian Georg).

De China officinarum. D. Lugduni Batavorum, 1715. 4°.

SCHWARZKOPF (S. A.).

Die narkotischen Genussmittel und die Gewuerze. I. Der Thee in naturhistorischer, diaetetischer, medicinischer und commercieller Hinsicht. II. Der Kaffee in naturhistorischer, . . . Hinsicht. Halle, 1881. 8°.

Der Kaffee in naturhistorischer, diaetetischer und medicinischer Hinsicht, seine Bestandtheile, Anwendung, Wirkung und Geschichte. Weimar, 1881. 8°.

SCHWEIZER (G.).

Der Futterbau. Eine prakt. Anleitung zum vortheilhaften Anbau saemmtl. Futterpflanzen, *etc.* Schaffhausen, 1875. 8°.

SCHWEITZER (Hermann).

Some remarks on Atropa Belladonna. *Pharm. Journ.* i. (1842) 517–520.

SCHWENCKE (Martin Wilhelm).

Kruidkundige beschrijving der in- en uitlandsche gewassen, welke heedendaagsche meest in gebruik zijn. 's Gravenhage, 1766. 8°.

SCHWILGUÉ (C. J. A.).

Traité de matière médicale. Paris, 1805. 2 vols. 12°. Ed. 2. 1809. Ed. 3. by P. H. Nysten, 1818. 2 vols. 8°.

SCOFFERN (John).

Manufacture of Sugar in the colonies and at home. London, 1849. 8°.

SCOTT (John).

Manual of Opium husbandry. Calcutta, 1877. 8°.

SCOTT (Wentworth Lascelles).

Report on the commercial powders of Ginger and Cinchona. *Pharm. Journ.* II. xi. (1869) 219–222.

SEABROOK (W. B.).

Memoir on the origin, cultivation, and uses of Cotton. Charleston, 1844. 8°.

SEATON (William John).

Report on the forests and alpha resources of Algeria. London, 1876. 8°.

SEBEÒK DE SZENT-MIKLOS (Alexander).

D. medico-botanica de Tataria hungarica. Viennae, 1779. 8°.

SEEMANN (Berthold).

On the Anta or Vegetable Ivory. [From Hooker's *Journ. Bot.*] *Pharm. Journ.* xi. (1851) 279–280.

On the Simara Cedron. [*Ib.*] *Pharm. Journ.* xi. (1851) 280.

On the Palo de Velas, or Candle tree. (Parmentiera cereifera, *Seem.*) [From Hooker's *Journ. Bot.*] *Pharm. Journ.* xi. (1851) 236.

SEEMANN (Berthold) *continued :—*

On the Jipijapa. (Carludovica palmata.) [From Hooker's *Journ. Bot.*] *Pharm. Journ.* xi. (1852) 365–366.

The Chinese Materia medica. [From Hooker's *Journ. Bot.*] *Pharm. Journ.* xi. (1852) 427.

On the cultivation of the Nutmeg at Singapore. [From Hooker's *Journ. Bot.*] *Pharm. Journ.* xi. (1852) 520.

On Gutta Taban. [From Hooker's *Journ. Bot.*] *Pharm. Journ.* xi. (1852) 575–576.

On Gambir. [Uncaria Gambir, *Roxb.* From Hooker's *Journ. Bot.*] *Pharm. Journ.* xi. (1852) 576–577.

On Nag-Kassar. [Calysaccion longifolium, *Wight.*] *Pharm. Journ.* xii. (1852) 62–63.

Remarks on Sarsaparilla. [Abstract from *The Gardeners' Chronicle.*] *Pharm. Journ.* xiii. (1854) 385.

Popular history of the Palms and their allies, containing a familiar account of their structure, geographical and geological distribution, history, properties and uses, and a complete list of all the species introduced into our gardens. London, 1856. 8°. Ed. 2. 1866.

—— Die Palmen . . . Unter Mitwirkung des Verfassers deutsch bearbeitet von Karl Bolle. Leipzig, 1857. 8°. Ed. 2. 1863.

On Anacahuite wood. [From *The Technologist.*] *Pharm. Journ.* II. iii. (1861) 164–165.

Report on the Tocuyo Estate of Venezuela. London, 1864. 8°.

SEMPLE (Robert Hunter).

Lecture on vegetable poisons, delivered at the establishment of the Pharmaceutical Society, June 22. *Pharm. Journ.* ii. (1842) 17–31.

Lecture on vegetable poisons, delivered Nov. 30, 1842. *Pharm. Journ.* ii. (1843) 429–442.

SÉNÉCLAUZE (Adrien).

Les conifères. Monographie descriptive et raisonnée, classée par ordre alphabétique de la collection compléte des conifères, tant indigènes qu'exotiques, *etc.* Paris, 1868. 8°.

SENFT (Ferdinand).

Lehrbuch der forstlichen Botanik. Jena, 1867. 8°.

Systematisch Bestimmungstafeln von Deutschlands wildwachsenden und cultivirten Holzgewaechsen, und den fuer sie wirklich schaedlichen Insectenarten. Ein Leitfaden auf Excursionen fuer Forstleute und alle Baumzuechter. Berlin, 1868. 8°.

SENGER (Gerhard Anton).

Die aelteste Urkunde der Papierfabrikation in der Natur entdeckt, nebst Vorschlaegen zu neuen Papierstoffen. Dortmund und Leipzig, 1799. 8°.

SERIALS. See *Periodical Publications* at end of Alphabetical list of Authors.

SERINGE (Nicolas Charles).

Description, culture et taille des mûriers, leurs espèces et leurs variétés. Paris, 1855. 8°. Atlas. 4°.

SERRANT (E.).

Les quinquinas en Algérie, étude sur leur acclimatation. Paris, 1879. 8°.

SERRES (Jean Joseph).

Description, culture et taille des mûriers, leurs espèces et leurs variétés. Paris, 1855. 8°.

SERVAIS (Gaspard Joseph de).

Korte verhandeling van de boomen, heesters en houtagtige kruid-gewassen, welke in de nederlandsche lugtstreek de winterkoude kunnen uitstaan, *etc.* Mechelen, 1789. 8°.

SESTINI (Fausto).

Il caffé; lettura fatta nell' instituto tecnico di Fochi. Firenze, 1868. 8°.

SESTINI (Fausto), & G. DEL TORRE.

Experiments on the cultivation of sugar-beet in the Campagna romana. *Chem. Soc. Journ.* xi. (1873) 1254–1255.

SETTEGAST (H.).

Analysis and use of lupine seed as fodder. *Chem. Soc. Journ.* x. (1872) 642–643.

SHARP (J. B.).

Observations upon the fibrous substances of the East and West Indies, and their applicability to supply the place of foreign productions for textile manufactures and paper. London, 1854. 8°.

Contains a lecture on Indian Fibres by J. F. Royle.

Tropical vegetable fibres. Address to the Chamber of Commerce, Dundee, Feb. 6, 1857. London, 1857. 8°.

SHAW (A. Nesbitt).

On the best means for promoting the growth and improving the quality of cotton in India. *Journ. Soc. Arts*, xi. (1863) 235.

SHELDRAKE (Timothy).

Botanicum medicinale, an herbal of medicinal plants, *etc.* London [1759]. fol.

SHERIFF (Daniel).

    The cultivation of Flax as a natural advantage, *etc.* [London, 1866.] 8°.

SHERWOOD (H.).

    An inquiry into vegetable fibres available for textile fabrics. [From *The Technologist.*] *Pharm. Journ.* II. viii. (1866) 81–84.

SHIER (John).

    Report on the Starch-producing plants of the Colony of British Guiana. Demerara, 1847. 8°.

SHORT (Thomas).

    A dissertation upon Tea; explaining its nature and properties. The natural history of Tea, *etc.* London, 1730. 4°.

    —— A dissertation upon Tea; explaining its nature and virtues, natural history, *etc.* Ed. 2. London, 1753. 4°.

    Medicina britannica, or a treatise on such physical plants as are generally to be found in the fields or gardens of Great Britain. London, 1747. 8°.

SHORTT (John).

    An essay on the culture and manufacture of Indigo, *etc.* Madras, 1860. 8°.

    —— An essay, *etc.* With a Hindustani translation. Madras, 1862. 8°.

    A handbook to Coffee planting in Southern India. Madras, 1864. 8°.

SHUTTLEWORTH (E. B.).

    Some experiments on the physiological effects of Coca. [Extract.] *Pharm. Journ·* III. viii. (1877) 221–222.

SICARD (Adrien Joseph Polyeucte).

    Monographie de la canne a sucre de la Chine, dite sorgho à sucre. Marseille, 1856. 8°. Ed. 2. 1858. 2 vols.

    Guide pratique de la culture du coton. Paris, 1866. 8°.

SIEBOLD (Philipp Franz von).

    Tabulae synopticae usus plantarum. In insula Dezima, 1827. fol.

    Synopsis plantarum oeconomicarum universi regni japonici. [Dezima, 1827.] 8°.

    Erwiederung auf W. H. de Vriese's Abhandlung: Het gezag van Kaempfer, Thunberg, Linnaeus en anderen, omtrent den botanischen oorsprong van den ster-anijs des handels, gehandhaafd tegen Dr. Ph. Fr. von Siebold en Prof. Zuccarini, *etc.* Leiden, 1837. 8°.

SIEGFRIED (J. J.).

Die Pflanzen in ihrer Anwendung auf Forst- und Landwirthschaft, Gartenbau, Gewerbe, und Handel. Zuerich, 1840.
8°.

SIEGMUND (Ferdinand).

Gemeinnuetziges Kraeuterbuch. Kurze Beschreibung aller als
Volkesheilmittel bekannten Pflanzen, *etc.* Wien, 1873–74.
8°.

Allgemeine illustrirte Kraeuterkunde und Volks-Arzenei Mittel-
lehre, *etc.* Bruenn, 1875. 8°.

SIERSTORPF (Caspar Heinrich, *Freiherr* von).

Ueber dei forstmaessige Erziehung, Erhaltung und Benutzung
der vorzueglichsten inlaendischen Holzarten. (Erster Theil.
1794. Die Forstbotanik, die Naturkunde der Baeume ueber-
haupt, und die Beschreibung der Eiche. Zweiter Theil, 1813.
Die Beschreibung der Fichte.) Hannover, 1794–1813. 2 vols.
4°.

SIEWART (Max).

Tanning materials of South America. [*Journ. Applied Science.*]
Pharm. Journ. III. viii. (1878) 548–549.

SIGMOND (George Gabriel).

Tea; its effects, medicinal and moral. London, 1839. 8°.

SILBERBERG (Louis).

Tobacco; its use and abuse. London, 1863. 8°.

SILLIMAN (Benjamin).

Manual on the Cultivation of the Sugar Cane, *etc.* Washington,
1833. 8°.

SILVA COUTINHO (Joao Martino da).

Gommes, résines, et gommes-résines. Exposition universelle de
1867 à Paris. [Rapports du jury international.] Paris,
1868. 8°.

SILVA LIMA (J. F. da).

Goa powder. (Translated and annotated by J. L. Paterson, M.D.)
[From *Med. Times and Gazette.*] *Pharm. Journ.* III. v. (1875)
723–725.

SILVESTRI (Antonio).

Le piante pratensi, ossia le erbe dei prati e dei pascoli Italiani,
descritte per famiglie naturali, *etc.* Torino, 1878. 8°.

SIMMONDS (Peter Lund).

Coffee as it is and as it ought to be. London, 1850. 18°.

On the medicinal properties of the Guaco. [Mikania Guaco.]
Pharm. Journ. x. (1851) 534–536.

SIMMONDS (Peter Lund) *continued :*—

The Commercial Products of the Vegetable Kingdom, considered in their various uses to man and in their relation to the arts and manufactures, forming a practical treatise and handbook of reference for the colonist, manufacturer, merchant, and consumer. London, 1853. 8°.

On some undeveloped and unappreciated articles of Raw Produce from different parts of the world. *Journ. Soc. Arts*, iii. (1854) 33–46.

On Agar-agar and Ceylon moss. *Pharm. Journ.* xiii. (1854) 355–356.

Vegetable wax from the Candleberry Myrtle. [Myrica cerifera.] *Pharm. Journ.* xiii. (1854) 418–423.

The medicinal plants of Australia. *Pharm. Journ.* xiii. (1854) 616–618.

The Gums and Resins of commerce. *Journ. Soc. Arts*, iv. (1855) 13–27.

A dictionary of trade products, commercial, manufacturing, *etc.* London, 1858. 8°. Ed. [2.] (A commercial dictionary of trade products), 1867.

On the culture and commerce of Chicory. *Pharm. Journ.* II. ii. (1860) 122–126.

Notes on Kittool fibre. [From *The Technologist.*] *Pharm. Journ.* II. iii. (1861) 280–281.

The trade in Liquorice. [The *Grocer.*] *Pharm. Journ.* II. iii. (1862) 623–625.

Coffee and Chicory, their culture, chemical composition, preparation for market, and consumption, *etc.* London, 1864. 12°.

Botany Bay, or Grass-tree Gum. [From *The Technologist.*] *Pharm. Journ.* II. viii. (1866) 78–80.

The Opium trade of China. *Pharm. Journ.* III. i. (1870) 361–362.

New Paper-making materials, and the progress of the Paper manufacture. *Journ. Soc. Arts*, xix. (1871) 171–179.

Saponaceous plants. [*Journ. Applied Science.*] *Pharm. Journ.* III. i. (1871) 585–586.

West Indian Medicinal plants. *Pharm. Journ.* III. i. (1871) 747.

The trade in Aloes. *Pharm. Journ.* III. iii. (1872) 83–84.

Indian Resins and Gums. [*Journ. Soc. Arts.*] *Pharm. Journ.* III. iii. (1872) 465–466.

On Nuts, their produce and uses. *Journ. Soc. Arts*, xx. (1872) 475–485.

SIMMONDS (Peter Lund) *continued* :—

Nuts, their produce and uses. [*Journ. Soc. Arts.*] *Pharm. Journ.* III. ii. (1872) 958–960, 977–980, 1018–1020, 1037–1038.

Supplies of Opium and Scammony from Turkey. *Pharm. Journ.* III. ii. (1872) 986–987, 1006–1008.

Fennel Flower Seed, or Black Cummin Seed. *Pharm. Journ.* III. iii. (1873) 684–685.

On the edible Starches of commerce, their production and consumption. *Journ. Soc. Arts*, xxi. (1873) 346–355. [Abstract in] *Pharm. Journ.* III. iii. (1873) 833–837, 853–855.

Wood products and Timber trade [of our Colonies]. *The Nautical Magazine*, xliii. (1874) 897–905, 1013–1019, xliv. (1875) 11–22. [*Anon.*]

Our Colonies: the vegetable Fibres they produce. *The Nautical Magazine*, xliv. (1875) 218–227. [*Anon.*]

Hops, their cultivation, commerce, and uses in various countries. A manual of reference, *etc.* London, 1877. 8°.

Tropical Agriculture. A treatise, *etc.* London, 1877. 8°.

The Commercial products of the Sea; or marine contributions to food, industry, and art. London, 1879. 8°.
> Various uses of seaweed, pp. 311-338.

SIMPSON (P. A.).

Native poisons of India. *Pharm. Journ.* III. ii. (1871) 604–606, 626–627, 665–667.

SINCLAIR (George).

Hortus gramineus Woburnensis, or an account of the results of experiments on the produce and nutritive qualities of different grasses and other plants used as the food of the more valuable domestic animals: instituted by John, Duke of Bedford. Illustrated with dried specimens of the plants upon which these experiments have been made. London, 1816. fol. Ed. 2. 1825. 8°. Ed. 3. 1826. Ed. 4. 1838. Ed. 5. 1869.

—— Johann Herzog von Bedford's Chemisch-agronomische Untersuchungen ueber den Werth verschiedner Futtergraeser. Zuerst herausgegeben von Sir Humphry Davy. Nach dem Franzoesischen von M. Marchais de Migneaux. Verdeutscht von A. A. Haas. Trier, 1821. 8°.

—— Hortus gramineus Woburnensis, oder Versuche ueber den Ertrag und die Nahrungskraefte verschiedener Graeser, *etc.* Aus dem Englischen von R. Schmidt. Stuttgart und Tuebingen, 1826. 4°.

Skene (—.).

Aleppo drugs. [Extract from Consular report.] *Pharm. Journ.*
III. iv. (1873) 189.

Slugg (Josiah Thomas).

The Pharmacy of the Bible. *Pharm. Journ.* III. ii. (1872)
804–805, 826–827.

Small (—.).

Cinchona cultivation in the Mauritius. [Extract from a Report.]
Pharm. Journ. III. vi. (1875) 21–22.

Smidth (Jens Hansen).

Arboretum scandinavicum. Fasc. 1. Kjoebenhavn, 1831. 12°.

Smith (—.).

Tsiology ; a discourse on Tea. Being an account of that exotic,
botanical, chymical, commercial and medical, with notices of
its adulteration, *etc.* London, 1826. 12°. [*Anon.*]

Smith (—.).

The Vegetable Ivory, or Tagua plant. (Phytelephas macrocarpa,
Willdenow.) *Pharm. Journ.* ii. (1843) 464.

Smith (Edward).

A glance at the Materia medica of Devon. *Pharm. Journ.* III.
viii. (1877) 215–218.

Smith (Frederick Porter).

The preparation and properties of the various kinds of Chinese
teas. [*Med. Times and Gazette.*] *Pharm. Journ.* III. ii.
(1871) 387–388, 405–406.

The oils of Chinese pharmacy and commerce. *Pharm. Journ.*
III. v. (1874) 61–62.

Smith (J. B.).

How can increased supplies of Cotton be obtained. *Journ. Soc.
Arts*, v. (1857) 374–386.

Smith (James Lawrence).

On the manufacture of the Otto of Rose at Kisanlik, in European
Turkey. *Pharm. Journ.* II. i. (1859) 143–144.

Smith (John), *Ex-Curator of Kew Gardens.*

Domestic Botany. An exposition of the Structure and Classification of plants, *etc.* London, 1871. 8°.

Smith (Junius).

Essays on the cultivation of the Tea-plant in the United States of
America. New York, 1848. 8°.

Smith (*Ricardo Carlos*) [Richard Charles].

Instrucções theoricas e practicas sobre a cultura do Holcus
saccharatus, ou cana doce de Imphée, e o seu producto em
aguardente, *etc.* Funchal, 1858. 8°.

SMITH (Walter George).

Note on Irish-grown Jalap. *Pharm. Journ.* II. xi. (1870) 842–844.

SMITT (J. W.).

Skandinaviens foernaemsta aetliga och giftige svampar. Stockholm, 1863. fol.

SMYTTÈRE (Philippe Joseph Emmanuel de).

Phytologie pharmaceutique et médicale ou végétaux envisagés sous les rapports anatomique, physiologique et thérapeutique. Paris, 1829. 8°.

Précis élémentaire de botanique médicale et de pharmacologie. Paris, 1837. 8°.

SOAMES (Peter).

A treatise on the manufacture of sugar from the sugar-cane. London, 1872. 8°.

SOLLY (Edward).

On Gutta Percha, a variety of Caoutchouc. *Pharm. Journ.* v. (1846) 510–514.

On recent adulterations of Opium. *Pharm. Journ.* vii. (1847) 104.

SONNINI DE MANONCOUR (Charles Sigisbert).

Mémoire sur la culture et les avantages du Chou-navet de Laponie. Paris, 1788. 8°.

Traité des Asclépiadées, particulièrement de l'Asclépiade de Syrie, procédé de quelques observations sur la culture du coton en France. Paris, 1810. 8°.

SOUBEIRAN (J. Léon).

Des applications de la botanique à la pharmacie. Thèse. Paris, 1854. 4°.

Note on a poisonous Loranthacea. [From *Journ. de Pharm.*] *Pharm. Journ.* II. i. (1860) 568–569.

Note on the Manna of Alhagi Maurorum, DC. [Extract.] *Pharm. Journ.* II. ii. (1861) 434.

Note sur la culture du cotonnier. Paris, 1865. 8°.

The collection of Mastic at Chios. [*Journ. de Pharm.*] *Pharm. Journ.* III. ii. (1871) 226.

Drugs from New Caledonia. [*Journ. de Pharm.*] *Pharm. Journ.* III. ii. (1871) 403–404.

SOUBEIRAN (J. Léon), & Auguste DELONDRE.

De l'acclimatation des Cinchonas dans les Indes néerlandaises et britanniques. Paris, 1867. 8°.

Note sur la culture des Cinchonas dans les indes Britanniques, et sur les échantillons d'écorces de cette provenance qui se trouvent à l'exposition de 1867. Paris, 1867. 8°.

Soubeiran (J. Léon) *continued* :—

Produits végétaux du Portugal considérés au point de vue de l'alimentation et de la matière médicale. Paris, 1867. 8°.

De l'introduction et de l'acclimatation des Cinchonas dans les Indes néerlandaisés et dans les indes Britanniques. Paris, 1868. 8°.

La matière médicale à l'exposition de 1867. Paris, 1868. 8°.

Sowerby (Leonard).

The Ladies Dispensatory: containing the natures, vertues, and qualities of all herbs and simples useful in physick, reduced into a methodicall order, *etc.* London, 1652. 8°.

Sowerby (William).

Catalogue of economic and medicinal plants growing in the gardens of the Royal Botanic Society of London. Part I. Stove and Greenhouse Plants. London, 1874. 8°.

Spall (P. W. A. van).

Verslag over de koffij- en kaneelkultuur op het eiland Ceijlon. Batavia, 1863. 8°.

Speck Obreen (H. A. van der).

Beschrijving van de timmerhoutsoorten die in Europeesch Guiana wassen. Rotterdam, 1864. 8°.

Speede (G. T. Frederick S. Barlow).

The new Indian Gardener, and guide to the successful culture of the kitchen and fruit garden in India, with illustrations. Also a vocabulary of the most useful terms . . . in the English and native . . . languages. Calcutta, 1848. 8°.

Spenner (Fridolin Carl Leopold).

Handbuch der angewandten Botanik, oder praktische Anleitung zur Kenntniss der medicinisch, technisch, und oekonomisch gebraeuchlichen Gewaechse Deutschlands und der Schweiz. Freiburg, 1834–36. 8°.

Spieler (Alexander Julius Theodor).

De plantis venenatis Silesiae. D. toxicologico-medica. Vratis-laviae, 1841. 8°.

Spiess (Ernst).

Wandtafel der wichtigsten Cultur- und Handelspflanzen. . . . Mit Text. Nuernberg, 1876. fol.

Spolverini (Giovan Battista).

La coltivazione del riso; o Il canapaio, di Girolamo Baruffaldi, nuovamente pubblicati a cura di G. Dehò. Torino, 1879. 32°.

Spon (E., & F. N.).

Encyclopaedia of the Industrial Arts, Manufactures, and Commercial Products. (Edited by C. G. Warnford Lock.) London, 1879, *etc.* 2 vols. 4°.—→

Camphor, pp. 571-578; Cane, 595-598; Chicory, 631-633; Cocoa, 684-691; Coffee, 691-722; Cork, 722-729; Drugs, 790-827; Dyestuffs, 854-869; Fibrous substances from plants, 909-1000; Fruit, 1022-1029; Narcotics, 1305-1351; Nuts, 1351-1360; Oils (vegetable, fixed, and volatile), 1377-1433.

The following will also be included: Resins and Gums; Spices; Starch; Sugar; Tannin; Tea; Timber; Wax.

Spon (Jacob).

Tractatus novi de potu Caphé, de Chinensium Thé, et Chocolata. Paris, 1865. 12°.

Drey neue Tractate von dem Trancke Cafe, Sinesischen Thé, und der Chocolata. Budissin, 1688. 8°. [*Anon.*]

J. S. Bevanda asiatica; hoc est, Physiologia potus café, *etc.* [Lipsiae,] 1705. 4°.

Spratt (George).

The medico-botanical Pocket-Book. London, 1836. 8°.

Spruce (Richard).

On the Siphonia, or India-rubber tree. [From Hooker's *Journ. Bot.*] *Pharm. Journ.* xi. (1857) 169–170.

Note on the India-rubber of the Amazon. *Pharm. Journ.* xv. (1855) 117–119.

Report on the expedition to procure seeds and plants of the Cinchona Succirubra, or red bark tree [in Ecuador]. London, 1861. 8°.

Notes on the cultivation of Cotton in Peru. London, 1864. 8°.

Spry (Henry Harpur).

Suggestions received by the Agricultural and Horticultural Society of India, for extending the cultivation and introduction of useful and ornamental plants, with a view to the improvement of the agricultural and commercial resources of India. Calcutta, 1841. 8°.

Squibb (Edward R.).

On commercial Jalap. [Extract.] *Pharm. Journ.* II. ix. (1868) 487–488.

Notes on Pareira. [*Amer. Journ. Pharm.*] *Pharm. Journ.* III. ii. (1872) 846.

Note on Aconite root. [*Proc. Amer. Pharm. Assoc.*] *Pharm. Journ.* III. iii. (1873) 974–976.

Notes on Aloes. [*Proc. Amer. Pharm. Assoc.*] *Pharm. Journ.* III. iii. (1873) 994–995.

SQUIER (Ephraim George).

    Tropical Fibres; their production and economic extraction. New York, 1861. 8°. *Also* London, 1863.

SQUIRE (Peter).

    Biennial Henbane. *Pharm. Journ.* III. xii. (1881) 256.

SQUIRE (William Stevens).

    On expressed juices [of officinal plants]. *Pharm. Journ.* i. (1841) 94–100.

    On the Daphne tribe of plants. *Pharm. Journ.* i. (1842) 395–399.

    Statement of the quantity of medicinal herbs supplied to the principal dealers in the London market. *Pharm. Journ.* v. (1846) 356.

STACEY (B. F.).

    The medicinal agents of the American Indians. [*Amer. Pharm. Assoc.*] *Pharm. Journ.* III. iv. (1874) 958–959.

STAHL's grosses illustrirtes Kraeuterbuch. Ausfuehrliche Beschreibung aller Pflanzen und Kraeuter in Bezug auf ihren Nutzen, ihre Wirkung und Anwendung, ihren Anbau, ihre Einsammlung und Aufbewahrung, *etc.* Neu-Ulm, 1877. 8°.

STAINBANK (II. E.).

    Coffee in Natal; its culture and preparation. London, 1874. 8°.

STANFORD (Edward Charles Cortis).

    On the manufacture of Kelp. *Pharm. Journ.* II. ii. (1862) 495–516.

STARK (James).

    Notice of the Copalchi-bark: a new and valuable bitter analogous to the Cascarilla. *Pharm. Journ.* ix. (1850) 463–466.

STECK (Abraham).

    De Sagu. D. Argentorati, 1757. 4°.

STECK (F. G.).

    De cultuur der Liberia Koffij. Batavia, 1879. 8°.

STEELE (James G.).

    Grindelia robusta. [Extract.] *Pharm. Journ.* III. vi. (1876) 566–568.

STEETZE (Adolph).

    On the Chaschisch of the Arabs (Cannabis sativa, or Hemp). [From the *Repert. f. d. Pharm.*] *Pharm. Journ.* v. (1845) 83.

STEIN (W.).

    On Wongshy, a new yellow dye. [From the *Pharm. Centr. Blatt.*] *Pharm. Journ.* ix. (1850) 541.

STEINMETZ (Andrew).

    Tobacco: its history, cultivation, manufacture, and adulterations. London, 1857. 8°. Ed. 2. 1857.

STELLA (Benedetto).

Il tabacco, . . . nella quale si tratta dell' origine, coltura, uso in fumo, in polvere, in foglia, in lambitivo et in medicina della pianta volgarmente detta tabacco, *etc.* Roma, 1669. 8°.

STENHOUSE (John).

On the dried Coffee-leaf of Sumatra, which is employed in that and some of the adjacent islands as a substitute for Tea or for the Coffee-bean. *Pharm. Journ.* xiii. (1854) 382–384.

Examination of Japanese Pepper, the fruit of the Xanthoxylum piperitum of De Candolle. *Pharm. Journ.* xvii. (1857) 19–21.

On some varieties of Tannin. [*Proc. Roy. Soc.*] *Pharm. Journ.* II. iii. (1861) 329–331.

On Wrightine; an alkaloid contained in the seeds of Wrightia antidysenterica. *Pharm. Journ.* II. v. (1864) 493–495.

STEPHENSON (John), & James Morss CHURCHILL.

Medical Botany : or, illustrations and descriptions of the medicinal plants of the London, Edinburgh, and Dublin Pharmacopoeas, *etc.* London, 1831. 4 vols. 8°.

STERBEECK (Francis van).

Citriculture, oft regeringhe der uythemsche boomen te weten oranien, citroenen, limoen, granaten, laurieren en andere. Waer in beschreven is de gedaente ende kennisse der boomen, met hunne bloemen, bladeren en vruchten, van ieder geslacht in het besonder. Als oock van den ranckappel, oprechten laurier van America, den caneelboom ende besonderlijck van den verboden Adams oft Paradysappel. T'Antwerpen, 1682. 4°. Ed. 2. 1712.

STERLER (Alois), & Johann Nepomuck MAYERHOFFER.

Europas medicinische Flora. Muenchen, 1820. fol.

STEVENS (James).

On Palm Sugar from India. *Pharm. Journ.* v. (1844) 65–66.

STEWART (F. L.).

Sorghum and its products. An account of recent investigations concerning the value of sorghum in sugar production, *etc.* Philadelphia, 1867. 8°.

Sugar made from Maize and Sorghum. A new discovery. New York, 18⁴0. 12°.

STEWART (John Lindsay), & Dietrich BRANDIS.

The Forest Flora of North-West and Central India. A handbook of the indigenous trees and shrubs of those countries. London, 1874. 8°.

STEWART (L. W.).

On the medicinal uses of the Indian species of Barberry. *Pharm. Journ.* II. vii. (1865) 303–305.

STILES (Matthew Henry).

The microscopic structure of the stem of Jaborandi (Pilocarpus species). *Pharm. Journ.* III. vii. (1877) 629–631.

STOCKHOLM (Jens).

Vejledning til en fordeelagtig Hoeravl, isaer for de Egue, hvor dens Dyrkning endnu er lidet bekjendt. Viborg, 1810. 8°.

STOCKS (J. Ellerton).

On two Balsam trees (Balsamodendra) from Scinde. [From Hooker's *Journ. Bot.*] *Pharm. Journ.* ix. (1849) 270–275.

STODDART (William Walter).

Bristol Pharmacology. *Pharm. Journ.* III. i. (1870) 482–483, (1871) 601–603, 661–663, 842–843, 881–883, 921–922, 985–986.

STOKES (Jonathan).

A botanical Materia medica, consisting of the generic and specific characters of the plants used in medicine and diet, with synonyms and references to medical authors. London, 1812. 4 vols. 8°.

STOUT (A. B.).

The Eucalyptus globulus in California. [Extract from a letter.] *Pharm. Journ.* III. iii. (1873) 603–604.

STRETTELL (George W.).

The Ficus elastica in Burma proper, or a narrative of my journey in search of it; a descriptive account of its habits of growth and the process followed by the Kakhyens in the preparation of caoutchouc, *etc.* Rangoon, 1876. 8°.

A new source of revenue for India. [The cultivation of fibre-yielding plants, especially Calotropis gigantea.] London, 1878. 8°.

STROBELBURGER (Johann Stephen).

Mastichologia, seu de universa mastiches natura dissertatio medica, in qua et arboris Lentisci et nobilissimi ejus gummi, resinae, mastiches, nomenclaturae, descriptiones, loci natales, culturae, qualitates, varietas, dignitas, electio, probatio, collectio, aestimatio, succedanea, utilitas ac inde parata medicamenta officinalia, magistralia, et artificialia describuntur. Lipsiae, 1628. 8°.

STRUMPF (Ferdinand Ludwig).

Die offizinellen Gewaechse in den natuerlichen Pflanzenfamilien, *etc.* Berlin, 1840. fol.

STUMPF (Johann Georg).

Praeses. De Robiniae Pseudoacaciae praestantia et cultu. Diss.
I. et. II. Gryphiae, 1796. 4°.

STUPPER (C. L.).

Medizinisch-pharmaceutische Botanik, oder Beschreibung und
Abbildung saemmtlicher in der neuesten k. k. oesterreichischen
Landes-Pharmakopoee vom Jahre 1836 aufgefuehrten Arznei-
pflanzen. Wien, 1841–43. 2 vols. 4°.

STURLER (W. L. de).

Catalogue descriptif des espèces de Bois de l'archipel des Indes-
Orientales . . . exposées à l'Exposition Internationale de 1867
à Paris. Leide, 1867. 8°.

SUAREZ DE RIVERA (Francisco).

Clave botanico, o medicina botanica, nueva y novissima. Madrid,
1738. 4°.

SUCKOW (George Adolph).

Oekonomische Botanik, zum Gebrauch seiner Vorlesungen.
Mannheim und Lautern, 1777. 8°.

Anfangsgruende der theoretischen und angewandten Botanik.
Leipzig, 1786. 3 vols. 8°. Ed. 2. 1797.

SUCKOW (Lorenz Johann Daniel).

Bezeichnung der vornehmsten Pflanzen und ihrer Kultur zum
Vortheile der Oekonomie. Ed. 4. Jena, 1794. 8°.

SULLIVAN (William K.).

On the manufacture of Beet-Sugar in Ireland. Dublin, 1851.
8°. *

Facts and theories; or, the real prospects of Beet Sugar manufacture in Ireland. Dublin, 1852. 8°.

SVENONIUS (Johannes).

Specimen de usu plantarum in Islandia indigenarum in arte
tinctoria. Hafniae, 1776. 8°.

SVERDRUP (Jacob).

Botanisk-oeconomisk Afhandling om Rugen. Christiania, 1823.
8°. Ed. 2. 1828.

SWAAB (S. L.).

Fibrous substances, indigenous and exotic, their nature, varieties,
and treatment considered, with a view to render them further
useful for textile and other purposes. London, 1864. 8°.

SWAYNE (George).

Gramina pascua, or a collection of specimens of the common pasture
grasses, with their Linnaean and English names, descriptions
and remarks, with 19 dried specimens. Bristol, 1790. fol.

SWINHOE (Robert).

Note on Formosa Camphor. [Extract.] *Pharm. Journ.* II. v. (1863) 280–281.

The Rice-paper of Formosa. *Pharm. Journ.* II. vi. (1864) 52–53.

Products of the island of Hainan. [From *The Field.*] *Pharm. Journ.* III. i. (1870) 529.

SYMES (Charles).

Carnauba root [Corypha cerifera]. *Pharm. Journ.* III. v. (1875) 661.

Baycuru [Statice sp. ?]. *Pharm. Journ.* III. ix. (1878) 196.

SYMONS (William).

Kauri Gum. *Pharm. Journ.* III. v. (1874) 259.

TAGAULT (Jean).

Tractatus de purgantibus medicamentis simplicibus. Basiliae, 1537. 4°.

TAMASSIA (Arrigo).

Atlante di tossicologia. Milano, [1879]. fol.

TARGIONI-TOZZETTI (Antonio).

Sommario di botanica medico-farmaceutica. Firenze, 1828. 8°.

Cenni storici sulla introduzione di varie piante nell' agricoltura ed orticultura toscana. Firenze, 1853. 8°.

TASSY (Louis).

Étude sur l'aménagement des fôrets. Paris, 1858. 8°. Ed. 2. 1872.

TASSY (V.).

Notes sur le pin Cembro. Digne, [1873]. 8°.

TATARINOW (Alexander).

Catalogus medicamentorum sinensium, quae Pekini comparanda et determinanda curavit. Petropoli, 1856. 8°.

TATHAM (William).

An historical and practical essay on the culture and commerce of Tobacco. London, 1800. 8°.

TAVANTI (Giuseppe).

Trattato teorico-pratico completo sull' ulivo, *etc.* Firenze, 1819. 2 vols. 8°.

TAYLER (William).

Thaumato-dendra, or the Wonders of Trees. *Journ. Soc. Arts,* xxv. (1877) 579–591.

TEICHMAYER (Hermann Friedrich).

Programma II. de Caapeba sive Parreira brava. Jenae, 1740. 4°.

TEMPLE (*Sir* Richard).
>Forest conservancy in India. *Journ. Soc. Arts*, xxix. (1881) 150–160.

TENNENT (*Sir* James Emerson).
>Notes on the Botany of Ceylon. [From Tennent's Ceylon; an account of the island, *etc.*, ed. 5, 1860.] *Pharm. Journ·* II. ii. (1860) 339–341.

TENORE (Michelo).
>Memorie sulle diverse specie e varietà di cotoni coltivate nel regno di Napoli, colle istruzzioni pel coltivamento del cotone siamese e le notizie sulle altre specie, di cui puossi provare l'introduzione. Napoli, 1839. 4°.

TERRACCIANO (Nicola).
>I legnami della Terra di Lavoro. Caserta, 1880. 8°.

TERZI (Ernesto).
>I prodotti delle conifere: memoria. Milano, 1872. 8°.

THAER (A.).
>Die landwirthschaftliche Unkraeuter. Farbige Abbildungen, Beschreibungen, und Vertilgungsmittel derselben. Berlin, 1881. 8°.

THIBAULT (Alexandre).
>Le China-grass, étude raisonnée de ce nouveau textile au point de vue de son acclimatation, de sa culture et de son emploi industriel. Du jute, du lin du Japon et autres textiles également propres à un emploi industriel. Nimes, 1866. 8°.

THIBIERGE (Adolphe), & —. REMILLY.
>De l'amidon du marron d'Inde: ou des fécules amylacées des végétaux non alimentaires aux points de vue économique, chimique, agricole et technique. Ed. 2. Paris, 1857.

THIÉBAUT-DE-BERNEAUD (Arsenne).
>Du genêt, consideré sous le rapport de ses différentes espèces, de ses propriétés et des avantages, qu'il offre à l'agriculture et à l'économie domestique. Paris, 1810. 8°.

THIELENS (Armand).
>Flore médicale belge. Bruxelles, 1862. 8°.

THIER (Léon de), & A. LEROY.
>Traité pratique de la culture des plantes fourragères. Bruxelles, 1865. 12°.

THIERRY DE MENONVILLE (Nicholas Joseph).
>Traité de la culture du nopal, et de l'éducation de la cochenille, dans les colonies françaises de l'Amérique. Précédé d'un voyage à Guaxaca. Au Cap Français, 1787. 2 vols. 8°.

THIRIAT (—.).

Rapport à la société d'agriculture du Gard sur la culture de la truffe. [Extract.] Nîmos, 1877. 8°.

Culture artificielle de la truffe. [Extract.] Nimes, 1879. 8°.

THOMAS (Jean Basile).

Traité général de statistique, culture, et exploitation des bois. Paris, 1840. 2 vols. 4°.

THOMSON (Anthony Todd).

Elements of Materia medica and Therapeutics, *etc.* London, 1832–33. 8°. Ed. 2. 1835. Ed. 3. 1843.

Introductory lecture on Materia medica, delivered at the establishment of the Pharmaceutical Society, Feb. 16. *Pharm. Journ.* i. (1842) 466–476.

Introductory lecture to a course of general and medical botany. Delivered at the establishment of the Pharmaceutical Society, May 11. *Pharm. Journ.* i. (1842) 620–632.

Synopsis of a course of Lectures on General and Medical Botany. *Pharm. Journ.* ii. (1843) 712–713 ; iii. (1844) 522–523.

Introductory Lecture to a course of Medical and General Botany. *Pharm. Journ.* ii. (1843) 755–766.

The Introductory Lecture on Botany. Delivered . . . on . . . April 24th. *Pharm. Journ.* iii. (1844) 561–565.

THOMSON (Thomas).

Root of Aconitum heterophyllum, *Wallich.* [Extract from a letter.] *Pharm. Journ.* xvi. (1856) 311–312.

THORNTON (Robert John).

A new family herbal : or popular account of the natures and properties of the various plants used in medicine, diet, and the arts. London, 1810. 8°. Ed. 2. 1814.

THOUIN (André).

Cours de culture et le naturalisation des végétaux. Paris, 1827. 3 vols. 8°.

THOZET (A.).

Notes on some of the roots, tubers, bulbs, and fruits used as food by the aboriginals of North Queensland. Rockhampton, 1866. 8°.

THURBER (George).

California Manna. [Extract.] *Pharm. Journ.* III. vii. (1877) 893.

THWAITES (George Henry Kendrick).

Cinchona cultivation in Ceylon. [Extract.] *Pharm. Journ.* II. v. (1864) 515–516.

TIDYMAN (Philipp).

Commentatio inauguralis do Oryza sativa. Goettingae, 1800. 4°.

TIEDEMANN (Friedrich).

Geschichte des Tabaks und anderer aehnlicher Genussmittel. Frankfurt-am-Main, 1854. 8°.

TIEMEROTH (Johann Heinrich).

Planta ac fructus Ananas hujusque usus medicus. D. Erfordiae, 1723. 4°.

TIMBAL-LAGRAVE (Édouard).

On the influence of cultivation upon medicinal plants. [Transl. from *L' Union pharmaceutique.*] *Pharm. Journ.* II. iii. (1862) 430–433.

Note pour servir à l'étude botanique et médicale de la valériane officielle. Toulouse, 1867. 8°.

TISON (Édouard).

Histoire de la fève de Calabar. Paris, 1873. 8°.

TODARO (Agostino).

Relazione sulla cultura dei cotoni in Italia, seguita da una monografia del genere Gossypium per servire di illustrazione alla raccolta dei cotoni presentata all' esposizione universale di Parigi nell' anno 1878. Roma e Palermo, 1879. 16°. Atlas fol.

TONNING (Henrik).

Norsk medicinisk och oekonomisk Flora, indeholdende adskillige Planter, som fornemmelig ere samlede i Tronheims stift. Forste deel [Linnean classes 1–15]. Kjoebenhavn, 1773. 4°.

TONNONI (P. A.).

Piante erbacea e legnose che più interessano l'agricoltura italiana, *etc.* Bologna, 1875. 8°.

TORELLI (Luigi).

L'Eucalyptus e l'agro Romano. Roma, 1878. 8°.

TORREY (John).

Notice of the "California Nutmeg." [Torreya californica, *Torr.*, From the *N. Y. Journ. Pharmacy.*] *Pharm. Journ.* xiv. (1854) 83–84.

TOURNEFORT (Joseph Pitton de).

Traité de la matiére médicale, ou l'histoire et l'usage des mèdicamens et leur analyse chimique. Ouvrage posthume de M. Tournefort, mis au jour par M. Besnier. Paris, 1717. 2 vols. 12°.

—— Materia medica, or a description of simple medicines generally us'd in physick. London, 1708. 8°. Ed. 2. 1716.

TRAPPEN (Jan Evert van der).

Specimen historico-medicum de Coffea, *etc.* Trajecti ad Rhenum, 1843. 8°.

TRATTINICK (Leopold).
Anleitung zur Cultur der aechter Baumwolle in Oesterreich.
Wien, 1797. 8°.

TRAUN (Heinrich).
Versuch einer Monographie des Kautschuks. D. Goettingen,
1859. 8°.

TRELOAR (Thomas).
The prince of Palms; being a short account of the Cocoa-nut tree,
showing the uses to which the various parts are applied, both
by the natives of India and Europeans. London, [1852]. 8°.

TRIANA (José).
Nouvelle Grenade ou états-unis de la Columbie. Catalogue de
l'exposition de M. José Triana. Paris, 1867. 8°.
Nouvelles études sur les Quinquinas, d'après les matériaux pré-
sentés eu 1867 à l'Exposition Universelle de Paris, et accom-
pagnée de fac-simile des dessins de la Quinologie de Mutis,
suivies de remarques sur la culture des quinquinas. Paris,
1870.. fol.
Gonolobus Cundurango. [*Comptes rendus.*] *Pharm. Journ.* III.
ii. (1872) 861.

TRIMEN (Henry).
The plants affording Myrrh. *Pharm. Journ.* III. ix. (1879)
893–894.

TROG (Jakob Gabriel).
Die essbaren, verdaechtigen, und giftigen Schwaemme der
Schweiz, *etc.* [Bern, 1845–50.] fol.
Die Schwaemme des Waldes als Nahrungsmittel, oder kurze
Anleitung zur Kenntniss der bei uns wildwachsenden essbaren
Schwaemme und zu ihrem zweckmaessigem Gebrauche. Bern,
1848. 8°.

TROJEL (Jakob Kofoed).
Priis-Skrift om Humle-Avlen. Kjoebenhavn, 1769. 8°. Ed. 2.
Odense, 1773.

TROUETTE (E.).
Acclimatation des arbres à caoutchoue à la Reunion. Paris, 1875.
8°.

TRUCHET (Michel de).
Mémoire sur la nécessité d'étendre la culture du tabac en France,
pour éviter l'exportation du numéraire; et sur l'examen analy-
tique des tabacs français, d'après lequel on peut avoir
l'assurance de trouver en eux les qualités nécessaires à la
bonne fabrication. Paris, 1816. 8°.

TUERK (Chr.).

Pflanzenkunde. Die wichtigsten wildwachsten Nutz- und Zierpflanzen mit Beruecksichtigung der vorzueglichten auslaend.
Gewaechese. Coburg, 1871. 2 vols. 8°.

TURNBULL (Alexander).

Ou the medical properties of the natural order Ranunculaceae,
and more particularly on the uses of Sabadilla seeds, Delphi-
nium Staphisagria, and Aconitum Napellus. London, 1835.
8°.

TURNER (J. A.).

The Cotton Planter's Manual: being a compilation of facts from
the best authorities on the culture of Cotton, its natural history,
chemical analysis, trade, and consumption, *etc.* New York,
1857. 8°.

—— Manual do plantador d'algodão. [Transl. by J. R. Jauffret.]
Maranhão, 1859. 4°.

TURNER (Robert).

Botanologia, the British physician; or the nature and vertues of
English plants, *etc.* London, 1664. 8°. [Ed. 2.] 1687.

TUSON (Richard V.).

Earth-nut, or Ground-nut, Cake. [*Veterinarian.*] *Pharm. Journ.*
III. vii. (1876) 332.

ULLGREN (Olof Matthias).

De plantis tinctoriis suecanis. D. I.-II. Upsaliae, 1815. 4°.

UMNEY (Charles).

Rhamnus Frangula versus Rhamnus catharticus. *Pharm. Journ.*
III. v. (1874) 21–22.

UNANUE (José Hipólito).

Disertacion sobre el aspecto, cultivo, comercio y virtudes de la
famosa planta del Perú, nombrada Coca, *etc.* Lima, 1794. 4°.

UNGER (Franz).

Botanische Streifzeuge auf dem Gebiete der Culturgeschichte.
I.-IX. Wien, 1857-68. 8°.

UNVERICHT (C.).

Betrachtungen und Rathschlaeger eines Botanikers in Beziehung
auf Waelder, Gehoelzpflanzungen aller Art und Schulgaerten,
etc. Karlowitz, 1874. 8°.

URE (Alexander).

On the Malambo, or Matias bark. *Pharm. Journ.* iii. (1843) 169–
170.

URE (Andrew).

An account of a substance recently imported from China, under
the name of Vegetable Wax, *etc.*   *Pharm. Journ.* vi. (1846)
69.

URICOECHEA (Ezequiel).

On the Cinchona barks of New Grenada.   *Pharm. Journ.* xiii.
(1854) 470–471.

USHER (Ralph).

English medicinal Rhubarb and Henbane.   [*Journ. Soc. Arts.*]
Pharm. Journ. II. ix. (1867) 81–86.

VALÉE (A.).

Culture de l'eucalyptus aux trois fontaines, près Rome.   Rome,
1879.   8°.

VALENTI SERINI (Francesco).

Dei funghi sospetti venenosi del territoria Senese.   Torino, 1868.
obl. 4°.

VALSERRES (Jacques).

Instructions pour la culture de la truffe.   Paris, 1876.   8°.

—— Question de reboisement.   Instructions pour la culture de la
truffe.   Ed. 2.   1877.   8°.

VANZINI (Carlo, & Luigi).

Istruzione pratica nella coltivazione del cotone adattato al clima
delle pianure dell' alta Italia.   Pallanza, 1863.   8°.

VANZO (L. D.)

Coltivazione dell' Elianto, memoria letta all' academia di Bovolenta
nella tornata 29 Ott. 1867.   Padova, 1868.   8°.

VASEY (George).

A catalogue of the Forest Trees of the United States which usually
attain a height of sixteen feet or more, with notes and brief
descriptions of the more important species, *etc.*   Washington,
1876.   8°.

VASSALLI-EANDI (Antonmaria).

Saggio teorico-pratico sopra l'Arachis hypogaea.   Torino, 1807.
8°.

VAUGHAN (James).

Notes upon the drugs observed at Aden, Arabia.   *Pharm. Journ.*
xii. (1852) 226–229, 268–271, (1853) 385–388.

On a bark called Heetoo, used in Abyssinia, with some remarks
on the Korarima and on Koussoo.   [With a note by D. Hanbury.]   *Pharm. Journ.* xii. (1853) 587–589.

VAUTHERIN (A.).

Des graines de Croton tiglium (petit pignon d'Inde) et de Curcas purgans (gros pignon d'Inde), des produits qu'on en retire, recherches de chimie, physique, physiologie, pharmacologie et micrographie. Paris, 1864. 8°.

VAVIN (Eugéne).

L'igname de Chine et son avenir. [Extrait.] Paris, 1878. 8°.

Igname ronde. [Extract.] Paris, 1879. 8°.

Rapport au nom de la commission de reboisement des montagnes par l'ailanthe. [Extract.] Paris, 1879. 8°.

Fenouil de Florence. [Extract.] Paris, 1879. 8°.

VAVASSEUR (P.), P. L. COTTEREAU, & A. GILLET DE GRANDMONT.

Dictionnaire universelle de botanique agricole, industrielle, mèdicale et usuelle, comprenant toutes le plantes vénéneuses et les champignons délétères et comestibles. Tome premier. Paris, 1836. 4°.

VEITH (Emmanuel).

Systematische Beschreibung der vorzueglichsten in Oesterreich wildwachsenden oder in Gaerten gewoehnlichen Arzneigewaechse, mit besondrer Ruecksicht auf die neue oesterreichsche Provinzial-Pharmacopoe. Wien und Trieste, 1813. 8°.

Abriss der Kraeuterkunde fuer Thieraerzte und Oekonomen, nebst einer Uebersicht der gewoehnlichsten einheimischen Gewaechse und ihrer Standoerter. Wien und Trieste, 1813. 8°.

VELLOZO (José Marianno da Conceição).

Quinografia portugueza, ou collecção de varias memorias sobre vinte e duas especies de quinas, tendentes do seu descobrimento nos vastos dominios do Brasil, copiada de varies authores modernos, enriquecida com cinco estampas de quinas verdadeiras, quatro de falsas, e cinco de balsameiras. Lisboa, 1799. 8°.

VERLOT (Bernard).

Rapport de la commission chargée de visiter les collections dendrologiques de Segrais. Paris, 1875. 8°.

VÉTILLART (Marcel).

Études sur les fibres végétales textiles, employèes dans l'industrie. Paris, 1876. 8°.

VETTORI (Piero).

Trattato di P. V. della coltivazione degli ulivi. Firenze, 1574. 8°.

VIBORG (Erik Nissen).

Efterretning om Sandvexterne och deres Anvendelse til at daempe Sandflugten paa Vesterkanten af Jyland. Kjobenhavn, 1788. 4°.

—— Beschreibung der Sandgewaechse, und ihrer Anwendung zur Hemmung der Flugsandes auf der Kueste von Juetland. Aus dem Daenischen, uebersetzt von J. Petersen. Kopenhagen, 1789. 8°.

Botanisk oekonomisk Afhandling om Bygget. Priisskrift. Kjobenhavn, 1788. 4°.

—— Botanisch-oekonomische Abhandlung von der Gerste. Priesschrift. Kopenhagen, 1802. 4°.

Botanisk Bestemmelse af de i danske Lov omtalte Sandvexter, somt om Sandflugtens Daempning. Kjoebenhavn, 1795. 8°.

Botanisk oekonomisk Beskrivelse over di i Landhuusholdningen vigtigte Aspe- og Pilearter. (Populus et Salix.) Et Priisskrift. Kjoebenhavn, 1800. 8°.

VICAT (Philipp Rudolfe).

Matière médicale, tirée de "Halleri Historia stirpium indigenarum Helvetiae, Bernae, 1768, folio," avec beaucoup d'additions. Berne, 1776. 2 vols. 8°. Ed. 2. (Histoire des plantes suisses, ou Matière médicale et de l'usage économique des plantes) 1791.

Histoire des plantes vénéneuses de la Suisse, contenant leur description, leurs mauvais effets sur les hommes et sur les animaux, avec leurs antidotes. Yverdun, 1776. 8°.

VIDAL (J.).

Conophallus Konjak, sa culture, ses usages comme plante alimentaire au Japon. [Extract.] Paris, 1877. 8°.

VIEILLARD (Eugéne).

Plantes utiles de la Nouvelle-Calédoine. [*Ann. Sci. Nat.* Ser. IV. tom. xiv.] Paris, 1862. 8°.

VIETZ (Ferdinand Bernhard).

Icones plantarum medico-oeconomico-technologicarum oder Abbildungen aller medizinisch-oekonomisch-technologischen Gewaechse, mit der Beschreibung ihres Nutzens und Gebrauches. Wien, 1800–20. 10 vols. 4°. Supplementum, 1822.

VIGNERON-JOUSSELANDIER (J. V.).

Manuel d'agriculture pratique des tropiques. Paris, 1860. 8°.

VIGUIER (L. G. Alexandre).

Histoire naturelle, médicale et économique des pavots et des Argemones. D. Montpellier, 1814. 4°.

Villa (Antonio, & G. B.):

I boschi nella Lombardia come prodotta di combustibile e di legname e como riparo a disastri meteorici, modo di rimettere i metesimi, di conservarli e di difenderli da guasti, massime degli insetti. Ed. 3. Milano, 1873. 8°.

Villa-Franca (—. de).

Note sur les plantes utiles du Brésil. [*Bull. de thérap. médicale,* 1879.] Paris, 1880. 8°.

Villeneuve (— de).

Traité complet sur la culture, fabrication et vente du tabac, *etc.* Paris, 1791. 8°. [*Anon.*]

Vincent (L.).

The Calabar bean. [*Journ. de Pharm.*] *Pharm. Journ.* III. ii. (1872) 906–907.

Vinen (E. Hart).

English oak galls. [*Journal of the Linnean Society.*] *Pharm. Journ.* xvi. (1856) 137–138.

Vinke (—.).

On Penghawar Djambi. [Extract.] *Pharm. Journ.* II. ii. (1860) 224–225.

Vinson (Jean François Dominique Émile).

Essai sur quelques plantes utiles de l'île Bourbon. Thèse. Paris, 1855. 4°.

Mémoire sur les essais d'acclimatation des arbres à quinquina, à l'île de la Réunion. Paris, 1875. 8°.

Violand (Adolphe).

Récolte des plantes mèdicinales. Colmar, 1862. 8°.

Virey (Julien Joseph).

Nouvelles considerations sur l'histoire et les effets hygiéniques du cafè, et sur le genre Coffea. Paris, 1816. 12°.

Monesia, or Buranhem of Brazil. [From the *Journ. de Pharmacie.*] *Pharm. Journ.* iv. (1844) 125–126.

Visiani (Roberto de).

Del metodo e delle avvertenze che si usano nell' orto botanico di Padova per la cultura, fecondazione, e fruttificazione delle Vaniglia. Memoria. Venezia, 1844. 4°.

Vittadini (Carlo).

Descrizione dei funghi mangerecci più comuni dell' Italia e de' velenosi che possono co' medisimi confondersi. Milano, 1835. 4°.

Vivenza (Andrea).

Mais; meliga, melgone, granturco, granone, ecc. Piacenza, 1881. 16°.

Viviani (Domenico).

I funghi d'Italia e principalmente le loro specie mangereccie, velenose e sospette, descritte ed illustrate con tavole disegnato e colorite dal vero. Fasc. 1–6. Genova, 1834–36. fol.

Voelcker (Augustus).

On the cultivation and uses of Sugar-Beet in England. *Journ. Soc. Arts*, xix. (1871) 307–318.

The Cultivation of the Beet-root in England. [*Journ. Soc. Arts.*] *Pharm. Journ.* III. i. (1871) 854–855.

Vogel (Heinrich).

Ueber die Cultur-Geschichte des Flachses und seinen Kampf mit der Baumwolle. Darmstadt, 1869. 8°.

Vogeli (Félix).

Flore fourragère, ou traité complet des alimens du cheval, *etc.* Paris, 1836. 8°.

Vogl (August E.).

Die Chinarinden des Wiener Grosshandels und der Wiener Sammlungen, mikroskopisch untersucht und beschrieben. Wien, 1867. 8°.

Ueber die Chinabaeume. Wien, *Schriften*, vii. (1868) 107–139.

Ueber vegetabilische Fette und Fettliefernde Pflanzen. Wien, *Schriften*, viii. (1869) 1–40.

Nahrungs- und Genussmittel aus dem Pflanzenreiche. Anleitung zum richtigen Erkennen und Pruefen der wichtigsten in Handel vorkommenden Nahrungsmittel, Genussmittel, und Gewuerze, mit Hilfe des Mikroskops. Zum allgemeinen sowie zum speci-ellen Gebrauche fuer Apotheker, Droguisten, Sanitaetsbeamte, *etc.* Wien, 1872. 8°.

—— Les aliments, guide pratique pour constater les falsifications des farines, fécules, cafés, chocolats, thés, épices, aromates, etc. Traduction par A. Focillon et G. Dauphin. Paris, [1875]. 16°.

The origin of the "Gum" of Quebracho colorado. *Pharm· Journ.* III. xi. (1880) 1–2.

Die gegenwaertig am haeufigsten vorkommenden Verfaelschungen und Verunreinungen des Mehles und deren Nachweisung. Wien, 1880. 8°.

Vriese. *See* De Vriese.

Vrij. *See* De Vrij.

Vrolik (Gerard).

Naamlyst der geneesryke plantgewassen in den Amsterdamschen Kruidtuin. Amsterdam, 1804. 8°.

V<small>ROLIK</small> (Gerard) *continued :*—

Catalogus plantarum medicinalium in Pharmacopoea batava memoratum. Editio altera auctior, *etc.* Amstelodami, 1805. 8°. Ed. 3. 1814.

V<small>RYDAG</small> Z<small>YNEN</small> (T.).

De in den handel voorkomende kinabasten, pharmacologisch behandeld. Rotterdam, 1835. 8°.

Chinae verae et Pseudo-Chinae herbarii regii Lugdunensis. Lugduni Batavorum, 1860. 4°.

W<small>ACKENRODER</small> (Heinrich).

On Coca leaves. [Transl. from *Archiv. der Pharm.*] *Pharm. Journ.* xiii. (1853) 224–225.

W<small>ADE</small> (Walter).

Salices, or an essay towards a general history of sallows, willows, and osiers, their uses and best methods of propagating and cultivating them. Dublin, 1820. 8°.

W<small>AECHTER</small> (J. K.).

Ueber die Reproductionskraft in Gewaechse insbesondre der Holzpflanzen. Ein Beitrag zur Pflanzenphysiologie mit Anwendung auf Forst- und Landwirthschaft und auf Garten-baukunst. Hannover, 1840. 8°.

W<small>AGNER</small> (A.).

Die Holzungen und Moore Schleswig-Holsteins. Hannover, 1875. 8°.

W<small>AGNER</small> (Daniel).

Pharmaceutisch-medizinische Botanik, oder Beschreibung und Abbildung aller in der k. k. Oestreich'schen Pharmacopoe vom Jahre 1820 vorkommenden Arzneipflanzen, in botanischer pharmaceutischer, medizinischer, historischer und chemischer Beziehung, *etc.* Wien, 1828. 2 vols. fol.

W<small>AHLBERG</small> (Goeran).

Om moegligheten att, enligt vegetabiliernas naturliga analogier, a priori bestaemma deras egenskaper och verkningar på meni-skliga organismen. D. I.–II. Upsala, 1834. 8°.

W<small>AITZ</small> (Frederik Augustus Karel).

Practische waarnmingen over eenige Javaansche geneesmiddelen, *etc.* Amsterdam, 1829. 8°.

——Praktische Beobachtungen ueber einige javanische Arnei-mittel . . . Aus dem Hollaendischen in das Deutsche uebersetzt und mit Anmerkungen begleitet von J. B. Fischer. Leipzig und Bruessel, 1829. 8°.

WALCHNER (Franz Hermann).

Darstellung der wigtigsten bis jetzt erkannten Verfaelschungen der Arzneimittel und Droguen, nebst einer Zusammenstellung derjenigen Arzneigewaechse, welche mit andern Pflanzen aus Betrug oder Unkenntniss verwechselt und in den Handel gebracht werden. Karlsruhe, 1842. 8°.

WALDSCHMIDT (Wilhelm Ulrich).

Programma de vegetabilium usu eximio in medicina. Kiliae, 1707. 4°.

WALKER (Campbell).

Reports on Forest management in Germany, Austria, and Great Britain . . . with extracts from reports by Mr. G. Mann, Mr. Ross, and Mr. T. W. Webber; and a memorandum by D. Brandis, . . . on the professional studies of forest officers on leave. London, 1873. 8°.

WALKER (William).

The Forests of British Guiana. *Proc. Royal Colonial Inst.* v. (1874) 126–156.

WALLACE (Alfred Russel).

Palm-trees of the Amazon, and their uses. London, 1853. 8°.

WALLIS (John).

Dendrology, in which are facts, experiments and observations demonstrating that trees and vegetables derive their nutriment independently of the earth. London, 1833. 8°.

WALPERS (Wilhelm Gerhard).

On White or Imperial Rhubarb. [Transl.] *Pharm. Journ.* xiii. (1855) 17–18.

WALTER (T. R. C.).

Notice of a poisonous leguminous plant from Swan River, New South Wales. *Pharm. Journ.* vi. (1847) 311–312.

WALTHER (Friedrich Ludwig).

Holzarten, die vorzueglichsten in- und auslaendischen. Bayreuth, 1790. 12°.

WALTON (Frederick).

On the introduction and use of elastic gums and analogous substances. *Journ. Soc. Arts,* x. (1862) 324–334.

WALZ (Isidor).

New method of distinguishing vegetable fibres. [*Journ. Applied Chem.*] *Pharm. Journ.* III. i. (1871) 749–750.

WANGENHEIM (Friedrich Adam Julius von).

Beschreibung einiger Nordamerikanischer Holz- und Buscharten, mit Anwendung auf deutsche Forste. Goettingen, 1787. 8°.

WANGENHEIM (Friedrich Adam Julius von) *continued :*—
Beitrag zur teutschen holzgerechten Forstwissenschaft, die Anpflanzung Nordamerikanischer Holzarten, mit Anwendung auf teutsche Forste, betreffend. Goettingen, 1787. fol.

WANKLYN (James Alfred).
Tea, Coffee, and Cocoa. A practical treatise on the analysis of Tea, Coffee, Cocoa, Chocolate, Mate, *etc.* London, 1874. 8°.

WARD (Ebenezer).
The Vineyards and Orchards of South Australia. (First Series) . . . being a series of articles . . . now reprinted. Adelaide, 1872. 8°.

WARD (H. Marshall).
Coffee Leaf disease. [Colombo, 1879.] fol.

WARD (J. S.).
Note on Australian opium. *Pharm. Journ.* III. i. (1871) 543, 553–554.

WARD (James).
Flax: its cultivation and preparation: with practical suggestions for its improvement and best mode of conversion. London, [1854]. 12°.

WARE (Lewis S.).
The Sugar Beet: including a history of the beet sugar industry in Europe, varieties, . . . soil, tillage, *etc.* Philadelphia, 1880. 8°.

WARING (Edward John).
On Hydnocarpus odoratus and Hydrocotyle asiatica as remedies in leprosy, scrophula, and secondary syphilis. *Pharm. Journ.* II. ii. (1860) 141–144.
Two remarkable Indian Fungi. [Extract from a letter.] *Pharm. Journ.* II. ii. (1861) 546–547.
On the seeds of Pharbitis Nil, *Choisy. Pharm. Journ.* II. vii. (1866) 496–499.
On the purgative action of certain Euphorbiaceous seeds. *Pharm. Journ.* II. vii. (1866) 550–560.
Pharmacopoeia of India, *etc.* London, 1868. 8°.
Remarks on the uses of some of the Bazaar medicines and common medical plants of India, *etc.* Ed. 2. London, 1874. 8°. Ed. 3. 1875. 16°.

WARMING (J. Eugene B.).
Teknisk-medicinisk-Botanik i Grundtraek, naermest som Erind-ringsond til Brug ved Forelaesninger. Den almindelige Del. Kjoebenhavn, 1877. 8°.

WARNES (John).

On the cultivation of Flax; the fattening of cattle with native produce, *etc.* London, 1846. 8°. Ed. 2. 1847.

Flax versus Cotton; or the two-edged sword against pauperism and slavery. No. 1. London, 1850. 8°.

No more published.

WARREN (Thomas T. P. Bruce).

On the cultivation of medicinal plants at Mitcham. *Pharm. Journ.* II. vi. (1864) 256–259.

The effects of soil and cultivation on the development of the active principles of plants. *Pharm. Journ.* II. vii. (1865) 210–216.

On the manufacture of India-Rubber, and its application to telegraphic purposes. *Journ. Soc. Arts,* xxvi. (1878) 128–143.

WASOWICZ (M. Dunin V.).

Aconitum heterophyllum, *Wall.* [*Archiv. der Pharm.*] *Pharm. Journ.* III. x. (1879) 301–303, 341–343, 463–464.

Japanese Aconite. *Pharm. Journ.* III. xi. (1880) 149–151.

WATSON (*Dr. —.*).

On the resin of Ceradia furcata. [From *Proc. Liverpool Lit. and Phil. Soc.*] *Pharm. Journ.* v. (1846) 366–369.

WATSON (John Forbes).

On the composition and relative value of the Food Grains of India. *Journ. Soc. Arts,* vi. (1857) 14–28.

On the growth of Cotton in India: its present state and future prospects, with special reference to supplies to Britain. *Journ. Soc. Arts,* vii. (1859) 278–292.

On the chief Fibre-yielding Plants of India. *Journ. Soc. Arts,* viii. (1860) 448–513.

The chief Fibre-yielding plants of India. London, 1860. 8°.

Index to the native and scientific names of Indian and other Eastern economic Plants and Products. London, 1866. 8°.

Report on the preparation and cultivation of Tobacco in India. London, 1871. fol.

Essay on the cultivation and manufacture of Tea. Calcutta, 1872. 12°.

The preparation and uses of Rhea Fibre. *Journ. Soc. Arts,* xxiii. (1875) 522–528.

Report on the preparation and use of Rheea fibre. London, 1875. fol.

Report on Cotton gins, and the cleaning and quality of Indian Cotton. London, 1879. fol.

WATSON (P. W.).

Dendrologia britannica, or trees and shrubs that will live in the open air of Britain throughout the year. A work useful to proprietors and possessors of estates, *etc.* London, 1825. 2 vols. 8°.

WATSON (Robert).

Notes on Maltese drugs. *Pharm. Journ.* III. viii. (1877) 341–342.·

WAUTERS (Pierre Engelbert).

Dissertatio botanico-medica de quibusdam plantis belgicis in locum exoticarum sufficiendis. Gandavii, 1785. 8°.

Repertorium remediorum indigenorum exoticis in medicina substituendorum. Responsum coronatum. Gandae, 1810. 8°.

WAYNE (Edward S.).

Oil of Cotton seed and Oil of Pignut Hickory. [*Proc. Amer. Pharm. Assoc.*] *Pharm. Journ.* xvi. (1856) 334–335.

Mata. [Extract.] *Pharm· Journ.* II. x. (1868) 27.

Cotton root [Gossypium herbaceum. From the *Amer. Journ. Pharm.*] *Pharm. Journ.* III. iii. (1872) 64–65.

WEATHERBY (William Henry).

Oleum Gossypii (Cotton-seed oil). [From *Amer. Journ. Pharm.*] *Pharm. Journ.* II. iii. (1861) 30–32.

WEBB (E. A.).

Remarks on a specimen of Chiretta. *Pharm. Journ.* III. i. (1871) 367–368.

False China root. *Pharm. Journ.* III. iii. (1873) 762–763.

WEBBER (Thomas Wingfield).

The Forests of England: their restoration and scientific management. *Journ. Soc. Arts,* xx. (1872) 215–222.

WEDDELL (Hugh Algernon).

Histoire naturelle des Quinquinas, ou monographie du genre Cinchona suivi d'une description du genre Cascarilla et de quelques autres plantes de la même tribu, *etc.* Paris, 1849. fol.

—— Naturgeschichte der Chinabaeume nebst einer Beschreibung der Genus Cascarilla und einiger anderer verwandter Pflanzen. In deutscher uebersetzung herausgegeben von Allgemeinen oesterreichischen Apotheker-Vereine. Wien, 1865. 8°.

The natural history of the Chinchonas. Extracted from a work on this subject by M. Weddell. [From the *Journ. de Pharm.*] *Pharm. Journ.* ix. (1849) 224–231.

On Cinchona Calisaya. [From the *Ann. des Sci. Nat.*] *Pharm. Journ.* ix. (1849) 232.

WEDDELL (Hugh Algernon) *continued* :—

> Table of commercial Cinchona barks, with the botanical species
> from which they are believed to be obtained. [From Weddell's
> *Hist. Nat. des Quinquinas.*] *Pharm. Journ.* ix. (1849) 240–
> 241.
>
> On the Cinchona barks. *Pharm. Journ.* ix. (1849) 267–270.
>
> On the Cephaelis Ipecacuanha. [From the *Répert. de Pharm.*]
> *Pharm. Journ.* ix. (1850) 332–333.
>
> Cinchona Calisaya. *Pharm. Journ.* ix. (1850) 365–369.
>
> Calisaya bark. *Pharm. Journ.* ix. (1850) 428–431.
>
> Trade in Cinchona bark in Bolivia. [Extract.] *Pharm. Journ.*
> xiv. (1854) 130–133.
>
> On the leaves of the Coca of Peru (Erythroxylon Coca, *Lamarck*).
> [Extract.] *Pharm. Journ.* xiv. (1854) 162–164, 213–215.
>
> On the extraction of Caoutchouc. *Pharm. Journ.* xv. (1855)
> 116–117.
>
> Native country of the Potato. [Extract.] *Pharm. Journ.* II. i.
> (1859) 339.
>
> Notes sur les Quinquinas. Paris, 1870. 8°.
>
> —— Uebersicht der Cinchonen. Deutsch bearbeitet von F. A.
> Flueckiger. Schaffhausen, 1870. 8°.
>
> —— Notes on the Quinquinas. London, 1871. 8°.

WEHNEN (—.).

> Bau, Leben, und Nahrungstoff der Culturpflanzen, *etc.* Berlin,
> 1881. 8°.

WEIN (E.).

> Die Sojabohne als Feldfrucht. Zusammenstellung der vorlie-
> genden Cultur- und Duengungsversuche fuer den praktischen
> Landwirth. Berlin, 1881. 8°.

WEINHOLD (Rudolph).

> Die wichtigsten wildwachsenden und angebauten Heil-, Nutz- und
> Giftpflanzen mit besondere Beruecksichtung der deutschen und
> schweizen Flora. Systematisch geordnet. Ein Huelfsbuch fuer
> Pharmaceuten, Kunst-, und Handelsgaertner, Landwirthe,
> Schullehre, *etc.* Bonn, 1871. 8°.

WEINRICH (Georg Albert).

> Dissertatio inauguralis de Haematoxylo campeachiano. Erlangae,
> 1780. 4°.

WEIR (John).

> On Myroxylon toluiferum, and the mode of procuring the Balsam
> of Tolu. [*Proc. Roy. Hort. Soc.*] *Pharm. Journ.* II. vi. (1864)
> 60–63.

Weise (Wilhelm).
　　Ertragstafeln fuer die Kiefer. Berlin, 1880. 8°.
Weiss (Eduard).
　　Der Hopfen. Betrachtet vom prakt. und wissenschaftlich Stand-
　　punkte. Nach besten Quellen, *etc.* Wien, 1878. 8°.
Weiss (Friedrich Wilhelm).
　　Entwurf einer Forstbotanik. Erster Band. Goettingen, 1775. 8°.
Weitenweber (Wilhelm Rudolph).
　　Der arabische Kaffe in naturhistorischer, diaetetischer und
　　medizinischer Hinsicht geschildert. Prag, 1835. 8°.
　　　　Contains, Literatura Coffeae, pp. 9-14.
Weitzner (Fr.).
　　Schulbotanik, oder Pflanzenkunde in Verbindung mit Techno-
　　logie. Breslau, 1853. 8°.
Wellcome (Henry S.).
　　A visit to the native Cinchona forests of South America. [*Proc.*
　　Amer. Pharm. Assoc.] *Pharm. Journ.* III. x. (1880) 980–982,
　　1000–1002, 1021–1022.
Welwitsch (Friedrich).
　　Synopse explicativa das amostras de madeiras e drogas medicinaes
　　. . . de Angola, *etc.* Lisboa, 1862. 8°.
　　Observations on the origin and geographical distribution of Gum
　　Copal in Angola. [From *The Gardeners' Chronicle.*] *Pharm.*
　　Journ. II. viii. (1866) 27–28.
Wencker (Daniel).
　　Dissertatio medica de potu Café, *etc.* [M. Mappus, *Praes.*]
　　Argentorati, [1693]. 4°.
Wenderoth (Georg Wilhelm Franz).
　　Dissertatio inauguralis medica, sistens materiae pharmaceuticae
　　hassiacae specimen. Marburgi, 1802. 8°.
　　Bemerkungen ueber wichtige einheimische Arzneipflanzen nebst
　　Vorschlaegen in Betreff derselben. (Akonitarzneien.) Kassel,
　　1837. 12°.
Wendt (Georg Friederich Carl).
　　Deutschlands Baumzucht, oder Verzeichniss der Holzarten,
　　welche das Klima von Deutschland im Freien aushalten, nebst
　　Angabe ihrer Groesse, des erforderlichen Bodens, Standes, der
　　Bluethezeit, Reife und Ausdauer. Eisenach, 1804. 4°.
Wendt (Johann Christian Wilhelm).
　　Anwiisning til et indsamle, torre og conservere de i Dannemark
　　og Norge vildvaxande mediciniske Planter. Kjoebenhavn,
　　1823. 8°.

WERNECK (Ludwig Friedrich Franz, *Freiherr* von).

Anleitung zur gemeinnuetzlichen Kenntniss der Holzpflanzen. Frankfurt-am-Main, 1791. 8°.

Versuch einer Pflanzenpathologie und Therapie. Ein Beitrag zur hoehern Forstwissenschaft. Mannheim und Heidelberg, 1807. 8°.

WEST (E.).

Emigration to British India, profitable investments for joint-stock companies, and for emigrants who possess capital. Ample supplies of raw Cotton, Silk, Sugar, Rice, Tobacco, Indigo, and other tropical productions. Employment for twenty millions of Hindu labourers upon one hundred million acres of fertile land in British India, which is now waste and unproductive. London, 1857. 8°. *

WESTRING (Johan Pehr).

Svenska lafarnas faerghistoria, eller saettet att anvaenda dem till faergning och annan hushållsnytta. Foerste bandet (Haeftet. 1–7). Stockholm, 1805. 8°.

—— Schwedens vorzueglichste Faerbeflechten treu nach der Natur abgebildet, nebst der chemischen Bearbeitung derselben, besonders in Ruecksight auf Faerberei. Aus dem Schwedischen uebersetzt von F. D. D. Ulrich. Norkoeping und Leipzig, 1805. 8°.

> Only the first fasciculus of the German edition appeared.

WHEELER (James Lowe).

Catalogus rationalis plantarum medicinalium in horto societatis pharmaceuticae Londinensis apud vicum Chelsea cultarum. Londini, 1830. 8°.

WHEELER (J. Talboys).

Hand-book to the Cotton Cultivation in the Madras Presidency, *etc.* Madras, 1863. 8°.

WHEELWRIGHT (W.).

Introductory remarks on the provinces of La Plata, and the cultivation of Cotton, *etc.* London, 1861. 8°.

WHISTLING (Christian Gottfried).

Oekonomische Pflanzenkunde fuer Lund- und Hauswirthe, Gaertner, Kuenstler, *etc.* Leipzig, 1805–7. 4 vols. 8°.

WHITE (H. A.).

Handbook to Cinchona planting for Ceylon planters. Colombo, 1877. 8°.

WHITE (Robert B.).

The Mikania Guaco as a remedy for snake bite. [Extract from a letter.] *Pharm· Journ.* III. xi. (1880) 369.

WHITEHEAD (Charles).

The cultivation of hops, fruit and vegetables. *Journ. Royal Agric. Soc.*, Scr. II. xiv. (1878) 723–760.

WHITEHOUSE (W. F.).

Letters and Essays on Agricola's Sugar Farming in Jamaica. Kingston, Jamaica, 1845. 8°.

WHITING (G.).

The products and resources of Tasmania, as illustrated in the International Exhibition, 1862. . . . Second edition, enlarged and corrected. With an appendix, containing papers on the Vegetable products exhibited by Tasmania, by . . . W. Archer; and on the Climate of Tasmania, by E. S. Hall. Hobart Town, 1862. 8°.

WIDDRINGTON (Samuel Edward), *formerly* COOK.

Spain and the Spaniards in 1843. London, 1844. 2 vols. 8°.

On the Forests of Spain. Oak, Pine, Ash. Vol. i. pp. 385-392.

WIESNER (Julius).

Die technisch von wendeten Gummiarten, Harze und Balsame. Ein Beitrag zur wissenschaftlichen Begruendung der technischen Waarenkunde. Erlangen, 1869. 8°.

Beitraege zur Kenntniss der indischen Faserpflanzen und der aus ihnen abgeschiedenen Fasern, nebst Beobachtungen ueber den feineren Bau der Bastzellen. Wien, 1870. 8°.

Die Rohstoffe des Pflanzenreiches. Versuch einer technischen Rohstofflehre des Pflanzenreiches. Leipzig, 1873. 8°.

Ueber die Bedeutung der technischen Rohstofflehre (technische Waarenkunde) als selbststaendiger Disciplin und ueber deren Behandlung als Lehrgegenstand an technischen Hochschulen. Dingler's *Polytech· Journ.* ccxxvii. (1880) 319–340.

WIESNER (T.).

Eucalyptus gum. [*Zeitschr. Apoth. Verein.*] *Pharm. Journ.* III. ii. (1871) 102–13.

WIGAND (Albert).

Der Baum. Betrachtungen ueber Gestalt und Lebensgeschichter der Holzgewaechse. Braunschweig, 1854. 8°.

WIGHT (Robert).

Notes on Cotton farming, explanatory of the American and East Indian methods. London and Reading, 1862. 8°.

WIGNER (G. W.).

On Tea. *Pharm. Journ.* III. vi. (1875) 261–263, 281–282, 402–404.

WILLEMOT (C.).

On the destruction of noxious insects by means of the Pyrethrum. [Pyrethrum Willemoti, *Duchartre.* Extract from *The Technologist.*] *Pharm. Journ.* II. v. (1863) 172–176.

WILLIAMS (D.).

On the farina of the Tacca pinnatifida. [From the *Journ. Agric. Hort. Soc. India.*] *Pharm. Journ.* vi. (1847) 383–384.

WILLKOMM (Moritz).

Deutschlands Laubhoelzer in Winter. Ein Beitrag zur Forstbotanik. Leipzig, 1859. 4°. Ed. 3. Dresden, 1880.

Die mikroskopischen Feinde des Waldes. Naturwissenschaftliche Beitraege zur Kenntniss der Baum- und Holzkrankheiten fuer Forstmaenner und Botaniker bearbeitet, *etc.* Dresden, 1866. 8°.

Forstliche Flora von Deutschland und Oesterreich oder forstbotanische und Pflanzengeographische Beschreibung aller im deutschen Reich und oesterreichische Kaiserstaat heimischen und im Freien angebauten Holzgewaechse. Nebst einem Anhang der forstliche Unkraeuter und Standortsgewaechse, *etc.* Leipzig, 1875. 8°.

Ueber europaeischer Culturpflanzen amerikanischer Herkunft. [Contained in *Sammlung gemeinnuetziger Vortraege* . . . No. 35.] Prag, 1877. 8°.

Waldbuechlein. Ein Vade mecum fuer Waldspaziergaenger. Leipzig, 1879. 8°. Ed. 2. 1880.

WILLS (George Sampson Valentine).

Manual of vegetable Materia medica, *etc.* London, 1877. 8°. Ed. 2. 1877. Ed. 3. 1878. Ed. 4. 1878. Ed. 5. 1880.

WILMER (Bradford).

Observations on the poisonous vegetables which are either indigenous in Great-Britain, or cultivated for ornament. London, 1781. 8°.

WILSON (John).

Recent improvements in the preparation and treatment of Flax. *Journ. Soc. Arts,* i. (1853) 229–231.

WINCKLER (Ferdinand Ludwig).

Die echten Chinarinden. Ein Beitrag zur genaueren Kenntniss dieser wichtigen Arzneimittel. Darmstadt und Leipzig, 1834. 4°.

WINCKLER (Ferdinand Ludwig) *continued :*—

On Maracaibo bark, a new kind of Cinchona bark, containing Quinoidine. [From the *Centr. Blatt.*] *Pharm. Journ.* x. (1851) 348.

WINKLER (Eduard).

Saemmtliche Giftgewaechse Deutschlands naturgetreu dargestellt und allgemein fasslich beschrieben, *etc.* Berlin, 1831. 8°. Ed. 3. 1853.

Abbildungen saemmtlicher Arzneigewaechse Deutschlands, welche in die Pharmacopoeen der groessern deutschen Staaten aufgenommen sind, *etc.* Leipzig, [1832, *etc.*]. 4°.

Handbuch der Gewaechskunde zum Selbststudium oder Beschreibung saemmtlicher pharmazeutisch-medizinischer Gewaechse, welche in den Pharmakopoeen der groesser deutschen Staaten aufgenommen sind. Leipzig, 1834. 8°.

Vollstaendiges Real-Lexicon der medizinisch-pharmaceutischen Naturgeschichte und Rohwaarenkunde, *etc.* Leipzig, 1840–42, 2 vols. 8°.

Abbildungen der Arzneigewaechse, welche homoeopathisch geprueft worden sind und angewendet werden. Leipzig, [1834–36]. 4°.

Pharmaceutische Waarenkunde, oder Handatlas des Pharmakologie, enthaltend Abbildungen aller wichtigen pharmaceutischen Naturalien und Rohwaren nebst genauer Charakteristik und kurzer Beschreibung. Leipzig, 1844–51. 8°.

Getreue Abbildung aller in der Pharmacopoeen Deutschlands aufgenommenen officinellen Gewaechse, nebst ausfuehrlicher Beschreibung derselben, in medicinischer, pharmaceutischer, und botanischer Hinsicht. Dritte verbesserte Auflage. Leipzig, [1846-47]. 4°.

Handbuch der medizinisch-pharmaceutischen Botanik. Nach den neuesten Entdeckungen bearbeitet. Leipzig, 1850. 8°.

WINSLOW (A. P.).

100 i ekonomiskt haenseende vigtiga svenska vaexter. Innhållende beskrifning på deras utseende, foernaemste egenskaper, anvaendling, odling . . . jemte bihang om potatis, sjukdomen, sot, brand, rost, och mjoeldryger. Goeteborg, 1871. 8°.

WINTER (L.).

On the cultivation of the Olive, near Ventimiglia. (From a letter addressed to Mr. Daniel Hanbury, F.R.S.) *Pharm. Journ.* III. iii. (1872) 182.

WINTHER (Paul).

Anviisning ag Underretning angaaende Tobaks-Plantning, grundet paa giorte Forsoeg. Kjoebenhavn, 1773. 8°. [*Anon.*]

WIRTGEN (Philipp).

Anleitung zur landwirthschaftlichen und technischen Pflanzen-kunde fuer Lehranstalten und zum Selbstunterricht. Erster und Zweiter Cursus. Coblenz, 1857-60. 8°.

WIRTH (Friedrich).

Der Hopfenbau. Eine gemeinfassliche belehr. Darstellung der Cultur und Behandlung des Hopfens von der ersten Anlage bis zur Ernte und dem Trocknen nach eigenegen Erfahrungen, *etc.* Stuttgart, 1875. 8°.

WISSETT (Robert).

A view of the rise, progress and present state of the Tea-trade in Europe. [London, 1801.] 8°.

Anonymous, unpaged, consisting of eleven sheets.

On the cultivation and preparation of Hemp; as also of an article produced in various parts of India called Sunn, which, with proper encouragement, may be introduced as a substitute for many uses to which hemp is at present exclusively applied. London, 1804. 4°.

WITTE (H.).

Flora. Afbeeldingen en beschrijvingen van boomen, heesters, éénjarige planten . . . voorkommenden in de Nederlandsche tuinen. Oorspronkelijke naar de natuur verwaardigde teeken-ingen van A. J. Wendel. Groningen, 1871. 8°.

WITTMACK (Louis).

Allgemeiner Katalog des Koeniglichen Landwirthschaftlichen Museums zu Berlin. Berlin, 1869. 8°. Ed. 2. 1873.

Fuehrer durch das Koenigl. Landwirthschaftliche Museum in Berlin. Berlin, 1873. 8°.

The fermentative action of the juice of the fruit of Carica Papaya. [Abstract.] *Pharm. Journ.* III. ix. (1878) 449-550.

WITTMANN (Karl).

Note on Sumbul. [Extract.] *Pharm. Journ.* III. vii. (1876) 329-330.

WODITSCHKA (Anton).

Die Giftgewaechse der oesterr.-ungarischen Alpenlaender und der Schweiz, mit besonder Beruecksichtung der Steiermark, *etc.* Ed. 2. Graz, 1874. 8°.

Wood (C. H.).

The process of Cinchona cultivation and alkaloid production in Bengal. *Pharm. Journ.* III. viii. (1878) 621–623. *See also* pp. 638–641.

Woodville (William).

Medical botany, containing systematic and general descriptions, with plates of all the medical plants comprehended in the catalogues of the Materia medica . . . of London and Edinburgh. London, 1790–93. 2 vols. 4°. Supplement, 1794. Ed. 2. 1810. Ed. 3. (in which 39 new plants have been introduced. The botanical descriptions arranged and corrected by William Jackson Hooker; the new medico-botanical portion supplied by G. Spratt) 1832. 5 vols.

Woolls (William).

Note sur la végétation, les produits et les charactères specifiques de quelques Eucalyptus d'après les travaux de William Woolls. Paris, 1876. 8°.

Wragge (A. Romaine).

Indian Forests and Indian Railways. *Journ. Soc. Arts*, xx. (1871) 79–91.

Wray (Leonard).

The Practical Sugar Planter; a complete account of the cultivation and manufacture of the sugar cane, according to the latest and most improved processes, *etc.* London, 1848. 8°.

The culture and preparation of Cotton in the United States of America, &c. *Journ. Soc. Arts*, vii. (1858) 77–89.

Tea, and its production in various countries. *Journ. Soc. Arts*, ix. (1861) 137–150.

On Indian Fibres. *Journ. Soc. Arts*, xvii. (1869) 452–459.

Wredow (Johann Christian Ludwig).

Oekonomisch-technische Flora Mecklenburgs, *etc.* Lueneburg, 1811–12. 2 vols. 4°.

Wright (Charles W.).

On Liquidambar styraciflua. [*Amer. Journ. Med. Science.*] *Pharm. Journ.* xvi. (1856) 326.

Wright (William).

A botanical and medical account of the Quassia Simaruba, or tree which produces the Cortex Simaruba. [Edinburgh, 1778.] 4°.

Wulfsberg (N.).

Holarrhena africana DC., eine tropische Apocynacee. [Inaugural Dissertation.] Goettingen, 1880. 8°.

WUNSCHMANN (Friedrich).

Deutschlands gefaehrliche Giftpflanzen naturgetreu dargestellt und nach ihren Wirkungen und Gegenmittel beschrieben, *etc.* Berlin, 1833. 8°.

YEATS (John).

Natural History of Commerce, *etc.* London, 1870. 8°.

YOUNG (Charles F. T.).

Flax, and improved machinery for its preparation. *Journ. Soc. Arts,* xv. (1867) 293–300.

YSABEAU (A.).

La science des campagnes. Plantes alimentaires et plantes four-ragères. Paris, 1862. 18°.

Végétaux cultivés. Plantes industrielles. Paris, 1862. 18°.

ZANON (Antonio).

Delle coltivazione e del uso delle patate e d'altre piante comesti-bili. Venezia, 1767. 8°.

ZENKER (Jonathan Carl).

Merkantilische Waarenkunde, oder Naturgeschichte der vorzueg-liehsten Handelsartikel, *etc.* Jena, 1831–35. 3 vols. 4°.

ZENNECK (Ludwig Heinrich).

Oekonomische Flora, oder systematisch tabellarische Beschreibung von 1000 fast ueberall in Deutschland wachsenden phanero-gamischen Pflanzen. Stuttgart und Prag, 1822. 4°.

ZIEGLER (Louis).

Die officinellen Gewaechse in tabellarischer Uebersicht nach dem kuenstlichen und natuerlichen Systeme geordnet. Hannover, 1845. 4°.

ZIGRA (Johann Hermann).

Dendrologisch-oekonomisch-technische Flora der im Russischen kaiserreiche bis jetzt bekannten Baeume und Straeucher, nebst deren vollstaendiger Kultur, *etc.* Dorpat, 1839. 2 vols. 8°.

   ·   In Russ, Petropoli, 1842.

ZIMMERMANN (J. H.).

Tabaksbaubuechlein. Kurze leichtfassliche Anleitung zur Pflanz-ung und Behandlung des Tabaks. Aarau, 1881. 8°.

ZIPPEL (Hermann), & Carl BOLLMANN.

Auslaendische Culturpflanzen in bunten Wandtafeln mit erlaeu-terndem Text. Braunschweig, 1876, *etc.* fol.—→

ZOELLER (T.).

Himalaya Tea. *Pharm. Journ.* III. ii. (1871) 162.

ZORN (Bartholomaeus).

Botanologia medica, seu delucida et brevis manuductio ad plantarum et stirpium tam patriarum quam exoticarum in officinis
pharmaceuticis usitatarum cognitionem, oder kurze Anweisung,
etc. Berlin, 1714. 4°.

ZORN (Johann).

Icones plantarum medicinalium. Abbildungen von Arzneigewaechse. Centuria I.–V. Nuernberg, 1779–84. 8°. Ed. 2.
Centuria I.–IV. 1784–90. [*Anon.*]

ZUCCARINI (Joseph Gerhard).

Charakteristik der deutschen Holzgewaechse im blattlosen Zustande, *etc.* Muenchen, 1823–31. 4°.

ZYNEN. *See* VRYDAG ZYNEN.

SERIALS.

GREAT BRITAIN AND IRELAND.

Belfast.

Flax Supply Association Reports. Annually since 1867. 8°.—>

Liverpool.

COPE's Tobacco Plant, a monthly periodical interesting to the manufacturer, the dealer, and the smoker. 1870, *etc.* fol.—>

London.

The Journal of Applied Science, a monthly newspaper devoted to art, agriculture, chemistry, *etc.* 1870, *etc.* fol.—> Edited by P. L. Simmonds.

SIMMONDS's [P. L.] Colonial Magazine and Foreign Miscellany. 15 vols. 1844–1848. 8°. *Continued as*

The Colonial Magazine and East India Review. Edited by W. H. G. Kingston. Vols. xvi.–xxiii. 1849–51. *United with the* Asiatic Journal, *and continued as*

The Colonial and Asiatic Review. 1849–51. 2 vols.

Pharmaceutical Journal and Transactions. 1842-59. 18 vols. 8°.

—— Second Series. 1860–70. 11 vols.

—— Third Series. 1870, *etc.*—>

Index to fifteen vols. 1857.

Index to twelve vols. . . . . . vol. xvi. Old Series to vol. ix. Second Series. 1869.

General Index to ten vols. . . . . . II. x. xi., III. i.–viii., July, 1868, to June, 1878. 1880. 4°.

The Journal of the Society for the Encouragement of Arts, Manufactures and Commerce. 1852, *etc.* 8°.—>

The Technologist; a monthly record of Science applied to Art and Manufacture. Edited by P. L. Simmonds. 1860, *etc.* 8°.—>

The Journal of Forestry, and Estates Management. 1877, *etc.* 8°.—>

Manchester.

The Sugar Cane, a Monthly Magazine devoted to the interests of the
Sugar Cane Industry. 1869, *etc.* 8°.—→

The Cotton Supply Reporter. 1858–72. 219 Nos. 4°.

Cotton. Continuation of the Cotton Supply Reporter. 1877-9. *

FRANCE.

Paris.

Société d'Acclimatation. Bulletin, 1854-63. 10 vols. 8°. II.
1864–73. 10 vols. III. 1874, *etc.*—→

Journal de l'Agriculture des Pays Chauds. Organe international du
progrès agricole . . . pour l'Algérie, les colonies françaises et
étrangères et la region intertropicale. Deuxième série. 1865,
etc. 8°. Edited by Paul Madinier.

> This is Série 2. of the Annales de l'Agriculture des Colonies.
> 1860-62. 4 vols.

GERMANY.

Berlin.

Forstliche Blaetter. Zeitschrift fuer Forst- und Jagdwesen. 1860–71.
12 ? vols. 8°. Edited by Julius Theodor Grunert.
—— Neue Folge. 1872, *etc.*—→ Ed. by J. T. Grunert &
Ottomar Vict.

Brunswick.

Deutscher Droguisten-Kalender. 1880, *etc.* 8°.—→ Edited by E.
Friese.

Leipzig.

Forstliche Berichte, *etc.* 1863–4. 8°.

Nordhausen.

Forstliche Berichte, mit Kritik ueber die neueste forstl. Journal-
Literatur. Neue Folge. 1852-61. 10 vols. *Continuation see*
Leipzig.

Osterode.

Forstliche Berichte, mit Kritik ueber die Hauptsaechlichste. der
Journal-Literature des Jahres 1842–51. 10 vols. 8°. Edited
by J. C. L. Schultze; *for continuation see* Nordhausen.

ITALY.

Sassari.

Giornale del laboratorio crittogamico et etnologico per lo studio dei
parassiti vegetali ed animali delle piante fanerogame della
Sardegna, diritto di Luigi Macchiati. 1879, *etc.* 4°.—>

INDIA.

Bombay.

Agri-Horticultural Society of Western India. Transactions. April
[-July], 1843. Nos. 1, 2. [1843]. 8°. Vol. i. 1852.
Agricultural and Horticultural Society of India. Transactions, 1837–
41. 8 vols. 8°.
    Monthly Journals, 1842–43. 2 vols. 8°. *Continued as*
    The Journal of the A. & H. Society of India. 1843, *etc.*—>

Calcutta.

Agricultural Gazette of India. 1869–74. fol. Edited by Robert
Knight. *
The Indian Forester, a quarterly magazine of forestry. Edited by
W. Schlich. 1876, *etc.* 8°.—>

Lahore.

Agri-Horticultural Society of the Punjab. Proceedings. 1852, *etc.*
8°.—>

ANONYMOUS PUBLICATIONS,

Exclusive of those works referred to their respective authors.

The nature of the drink Kauhi, or Coffee, and the berry of which it is made, described by an Arabian phisitian. Oxford, 1659. 8°.

> This is a singular little tract of two pages in Arabic, with an English translation facing ; on the title-page in the British Museum copy (pressmark 546 b. 24), has been written in an old hand 'Dr. Pococke yᵉ Translator.

Natural History of the Coffee, Thee, Chocolate, Tobacco . . . with a tract of Elder and Juniper-berries . . . and also the way of making Mum, *etc.* London, 1682. 4°.

—— Natur- gemaessige Beschreibung der Coffee, *etc.* Hamburg, 1684. 12°.

The present state of the Sugar plantations considered; but more especially that of the island of Barbadoes. London, 1713. 8°.

Remarks upon a book, intituled, "The present state of the sugar colonies considered, etc." London, 1731. 8°.

Lettre à M. Le Monnier, de l'Académie des Sciences, premier médecin ordinaire du Roi, sur la culture du Café. Amsterdam, 1773. 12°. *

The history of the tea plant; from the sowing of the seed, to its package for the European market, including every interesting particular of this admired exotic. To which are added, remarks on imitation tea, extent of the fraud, legal enactments against it, and the best means of detection. London, [1820 ?] 8°.

Papers respecting the culture and manufacture of Sugar in British India ; also notices of the cultivation of Sugar in other parts of Asia ; with miscellaneous information respecting Sugar. London, 1822. fol. *

Lehrbuch der forst- und landwirthschaftlichen Naturkunde. Tuebingen, 1827–40. 4 vols. 8°.

A description and history of vegetable substances used in the arts, and
in domestic economy. Timber Trees, Fruits. London, 1829. 8°.

> Forms vol. ii. of the Library of Entertaining Knowledge.

Vegetable substances. Materials of Manufactures. London, 1832. 8°.

> Forms vol. xv. of the Library of Entertaining Knowledge.

Reports and Documents connected with the proceedings of the East
India Company, in regard to the culture and manufacture of
Cotton-Wool, Raw Silk, and Indigo, in India. London, 1836. 8°.

Cotton, from the Pod to the Factory, a popular view of the natural
and domestic history of the plant . . . with the rise and progress
of the Cotton Factory, *etc.* London, 1842. 8°.

> (New Library of Useful Knowledge.)

The Indian Hemp, or Gunjah (Cannabis indica). *Pharm. Journ.* i.
(1842) 489–490.

Indigenous plants, the roots, rhizomes, cormi, leaves or flowers of
which may be gathered in July, arranged according to the
Natural Orders. *Pharm. Journ.* ii. (1842) 35–38.

Medicinal plants, which should be collected in August, arranged in
the Natural Orders. *Pharm. Journ.* ii. (1842) 73–77.

Eight practical treatises on the cultivation of the sugar cane, *etc.*
Jamaica, 1843. 8°.

> The authors' names are given as: I. Thomas Henry; II. Raynes
> W. Smith; III. Anonymous; IV. W. F. Whitehouse; V. James
> Sullivan; VI. W. A. Clements; VII. G. W. Gordon; VIII.
> Anonymous.

Plants which may be gathered, or their roots dug up in the month of
May. *Pharm. Journ.* ii. (1843) 721.

On Patchouli, or Puchá Pát. *Pharm. Journ.* iv. (1844) 80–82.

On Chicory, or wild succory. *Pharm. Journ.* iv. (1844) 119–121.

> This and the preceding article are probably by Dr. Pereira.

Report of a committee appointed for the examination of a new mode
of cultivating the sugar-canes, introduced by Mr. V. Gallet.
Trans. Royal Soc. Mauritius, vol. i. pt. 2 (1845) 123–135.

> The text is in French.

Clove Bark. [Laurineae.] *Journ. Pharm.* iv. (1845) 466.

Luffa purgans and Luffa drastica. South American colocynth. [Cf.
SCOTT.] *Pharm. Journ.* iv. (1845) 466.

On the Momordica-bucha, commonly called Cabacinho, or Bucha of
the hunters of Pernambuco. [Luffa purgans, *Mart.*] *Pharm.
Journ.* v. (1846) 569–570.

Poisoning with the berries of Atropa Belladonna, or Deadly Night-shade. [With a woodcut and description from *The Gardeners' Chronicle*.] *Pharm. Journ.* vi. (1846) 174–177.

The varieties of the Almond. *Pharm. Journ.* vi. (1846) 222–224.

Of the tree which yields the Gutta Percha or Gutta Turban. [From the *Journ. Agri. Hort. Soc. India,* and *Lond. Journ. Bot.*] *Pharm· Journ.* vi. (1847) 379–381.

On Patchouli. [Extracts.] *Pharm. Journ.* vi. (1847) 432. *See also* vol. iv. 80.

On Korarima. *Pharm. Journ.* vii. (1847) 16. *See also* BEKE.

On Sumbul Root. *Pharm. Journ.* vii. (1848) 546.

On the Revalenta arabica. *Pharm. Journ.* viii. (1848) 30. *See also* vol. iv. (1845) 415.

On the cultivation of Saffron in France and Austria. *Pharm. Journ.* viii. (1848), 171–173.

Report on the Growth of Cotton in India. London, 1848. fol. *

Evidence of the Committee on the British West Indies, and foreign [sugar-]growing Countries. London, 1848. 8°. *

Experiments on Cotton in the Southern Mahratta Country, by order of the East India Company, from 1830–48. Bombay, 1849. fol. *

Mecca or Bussorah Galls. *Pharm. Journ.* viii. (1849) 422–424.

Umbilicus pendulinus (De Candolle), Cotyledon Umbilicus (Linn.). Navelwort or Wall Pennywort. [With a quotation from a paper by Salter, from the *Medical Gazette*.] *Pharm. Journ.* viii. (1849) 526–527.

Deïama, or Tobacco of Congo. [From the *Journ. de Pharm.*] *Pharm. Journ.* ix. (1849) 143.

Lentils, Revalenta, Ervalenta. *Pharm. Journ.* x. (1850) 64–66.

Mitcham; its physio gardeners and medicinal plants. *Pharm. Journ.* x. (1850) 115–119, 168–172, 236–239, 297–299, (1851) 340–342.

Chinese Galls. *Pharm. Journ.* x. (1850) 127–129.

On the Aegle Marmelos, or Indian Bael. *Pharm. Journ.* x. (1850) 165–167.

Babul bark. [Acacia arabica.] *Pharm. Journ.* x. (1850) 177.

A Descriptive and Historical Account of the Cotton Manufacture of Dacca, in Bengal. By a Former Resident of Dacca. London, 1851. 8°. *

Coffee and its adulterations. [From the *Lancet*.] *Pharm. Journ.* x. (1851) 394–396.

Exhibition of Industry, 1851. [Raw materials and produce.] 1. British Guiana; 2. Barbadoes. *Pharm. Journ.* x. (1851) 407–412.

Arctopus root (Radix Arctopi echinati). *Pharm. Journ.* x. (1851) 559.

Great Exhibition of the Works of Industry of all Nations. Chemical and Pharmaceutical Products. *Pharm. Journ.* xi. (1851) 8–21, 68–73, 107–114, 155–161, 223–228, 263–265.

Cotyledon orbiculata. *Pharm. Journ.* xi. (1851) 106–107.

Cultivation and manufacture of Patna Opium. *Pharm. Journ.* xi. (1851) 205–212.

Monopoly of Yellow or Calisaya Bark. *Pharm. Journ.* xi. (1851) 215–218.

New Lebanon; its physic gardens and their products. *Pharm. Journ.* xi. (1852) 323–325, 413–414.

Note on the oil of Geranium having the odour of Roses (Pelargonium odoratissimum). *Pharm. Journ.* xi. (1852) 325.

Stillingia sebifera, or Tallow tree, and vegetable tallow of China. [From Hooker's *Journ. Bot.*] *Pharm. Journ.* xii. (1852) 73–74.

Chinese Wax, Pe-la, or insect-wax. [From Hooker's *Journ. Bot.*] *Pharm. Journ.* xii. (1852) 74–75.

Wood wool. [From *The Gardeners' Chronicle.*] *Pharm. Journ.* xii. (1853) 430–431.

Use of Coffee-leaves in Sumatra. *Pharm. Journ.* xii. (1853) 443–444.

Bush Tea of the Cape of Good Hope. *Pharm. Journ.* xiii. (1853) 172.

Importation of Wurrus (Capila-podie dye of Ainslie). *Pharm. Journ.* xiii. (1853) 284.

Siberian Rhubarb. *Pharm. Journ.* xiii. (1854), 329.

Guaco, an antidote for Snake bites. *Pharm. Journ.* xiii. (1854) 412–413.

Hydrocotyle asiatica, a cure for Leprosy. [From the *Madras Gazette.*] *Pharm. Journ.* xiii. (1854) 427–429.

Exposition universelle de 1855. Catalogue des produits naturels, industriels, artistiques préséntés par le royaume de Sardaigne, précéde d'une Introduction sur les produits et sur les principales industries des états sardes et avec notes explicatives. Paris, [1855]. 8°.

Purgative fruit from South America [Luffa sp. ?]. *Pharm. Journ.* xiv. (1855) 300–301.

Chironia chilensis, *Willd.*, Canchalagua. *Pharm. Journ.* xiv. (1855) 326.

Notes upon the Materia medica of Scinde. *Pharm. Journ.* xiv. (1855) 456–462.

Notes upon the Materia medica of the Paris Exhibition, 1855. *Pharm. Journ.* xv. (1855) 267–269; (1856) 332–335, 342–350.

Madras Exhibition of Raw Products, Arts, and Manfactures of Southern India, 1855. Reports by the Juries, *etc.* Madras, 1856. 4°.

Reports of the Madras Government on the Fibres of Southern India. Madras, 1856. 8°. *

Mammoth tree of California. *Pharm. Journ.* xv. (1856) 519.

Florida Indigo. [From the *Florida News.*] *Pharm. Journ.* xvi. (1856) 337.

Introduction of the culture of Cinchona into Java. [From *Bonplandia.*] *Pharm. Journ.* xv. (1856) 471.

Barbadoes farming. Hints to young Barbadoes planters. Barbadoes, 1857. 8°. *

The cultivation of Orleans staple Cotton, from the improved Mexican seed, as practised in the Mississippi cotton growing region. Manchester, 1857. 8°.

Siam Cardamoms. *Pharm. Journ.* xvi. (1857) 556.

Destruction of the Gutta Percha trees in Singapore. [From Hooker's *Journ. Bot.*] *Pharm. Journ.* xvii. (1857) 193–194.

Hydrocotyle asiatica. *Pharm. Journ.* xvii. (1857) 312.

Sanguinaria canadensis, *Linn.* *Pharm. Journ.* xvii. (1857) 312–313.

Madras Exhibition of Raw Products, Arts, and Manufactures of Southern India, 1857. Reports by the Juries, *etc.* Madras, 1858. 4°.

African Senna. *Pharm. Journ.* xvii. (1858) 499.

Madras Exhibition of 1859 of the Raw Products of Southern India. On the Cotton of the gigantic swallow wort, (Calotropis gigantea); as, also, on the silk worm and silk manufacture in Bengal, Bombay, China, Madras, and Mysore. Madras, 1858. 8°.

> Pp. 158-282 consist of a Dissertation on the silk manufacture and the cultivation of the mulberry, translated from the works of Tseu-Kwang-K'he, . . Shanghae, 1849.

Introduction of the Cinchona into India. *Pharm. Journ.* II. i. (1859) 29.

Note on the tree upon which the insect producing the white wax of China feeds [Fraxinus chinensis, *Roxb.*]. *Pharm. Journ.* II. i. (1859) 176.

Note on Japan wax. *Pharm. Journ.* II. i. (1859) 176–178.

Veratrum viride. American Hellebore. *Pharm. Journ.* II. i. (1859) 186–187.

Manufacture of Otto of Roses. *Pharm. Journ.* II. i. (1859) 264–267.

The Tea plant, its history and uses. London, 1860. 16°.

Papers on the cultivation of Indigo in Bengal. Report of the Indigo Commission. Calcutta, 1860. 8°. *

Sale of Dandelion Coffee. *Pharm. Journ.* II. i. (1860) 346–348, 357–358, 396.

Introduction of the Tea plant into the United States. [From *The Gardeners' Chronicle.*] *Pharm. Journ.* II. i. (1860) 429–430.

Alleged adulteration of Pepper. *Pharm. Journ.* II. i. (1860) 534–535.

On the introduction of the Cinchona trees into India. *Pharm. Journ.* II. ii. (1860) 201–204.

Collection of Mastic at Chios. [Extract.] *Pharm. Journ.* II. ii. (1860) 282.

Yellow Poppy seed oil. [From *The Technologist.*] *Pharm. Journ.* II. ii. (1860) 283.

Indigo, and its enemies; or facts on both sides. By Delta. Ed. 3. London, 1861. 8°.

Notes upon the Cocoa-nut tree and its uses. Edinburgh, 1861. 8°. *

The Cotton supply. A letter to John Cheetham, Esq., by a Fellow of the Royal Geographical Society. London, 1861. 8°.

El cultivo del Algodon llamado nueva orleans, producido de la semilla mejicana mejorada del modo practicado en le region algodonera del Mississippi. Manchester, 1861. 8°.

Vegetable silk and wool. London, 1861. 8°.

Aegle Marmelos, or Indian Bael. *Pharm. Journ.* II. ii. (1861) 499–500.

Chicory, the quantity produced, and the duty charged on it. *Pharm. Journ.* II. iii. (1861) 183–184.

The root of Ginseng. [Extract.] *Pharm. Journ.* II. iii. (1861) 197.

The Cinchonae in India. [Extract.] *Pharm. Journ.* II. iii. (1861) 217.

Cotton culture in new or partially developed sources of supply. Report of proceedings at a conference, Aug. 13, 1862. Manchester, 1862. 8°.

List of specimens of some of the woods of British Burmah, sent to England for the International Exhibition of 1862. Rangoon, 1862. 4°.

Report of the Commissioners of Inland Revenue. [On Tobacco, Snuff, Pepper, Coffee, Tea, Beer and Hops, *etc.*] *Pharm. Journ.* II. iv. (1862) 164–171.

Pharmaceutical products in the Exhibition. Unmanufactured drugs of vegetable origin. *Pharm. Journ.* II. iv. (1862) 219–221.

Le cotonnier, son histoire, sa culture en Algerie. Marennes, 1863. 8°. *

Introduction of Cinchona into India. *Pharm. Journ.* II. v. (1863) 105–109, 156–159.

Cortex Musenae. [Extract.] *Pharm. Journ.* II. v. (1863) 184.

Relazioni dei giurati della prima exposizione dei cotoni italiani. Torino, 1864. 8°. *

Question cotonnière. Paris, 1864. 8°. *

Cinchona cultivation in Jamaica. [*Jamaica Guardian.*] *Pharm. Journ.* II. v. (1864) 512–515.

Tea cultivation [in India. Calcutta, 1865.] 8°.

Uses of the Horse-Chestnut. [From *The Technologist.*] *Pharm. Journ.* II. vi. (1865) 486–488.

A new oil-seed for the colonies. [Madia sativa. From *The Technologist.*] *Pharm. Journ.* II. vii. (1865) 33.

The Flower farms of France. [From *The Amer. Gardener's Monthly.*] *Pharm. Journ.* II. vii. (1865) 291–293.

Culture et production de coton dans les colonies françaises. Paris, 1866. 8°. *

Handatlas saemmtlichen medicinisch-pharmaceutischen Gewaechse oder naturgetreue Abbildungen und Beschreibungen der officinellen Pflanzen, *etc.* Jena, 1866–70. 8°.

Copal resin. [From *The Gardeners' Chronicle.*] *Pharm. Journ.* II. vii. (1866) 424–425.

Port-royal Senna. *Pharm. Journ.* II. vii. (1866) 447–448.

New paper material. [The roots of Lucerne, and other species of Medicago. From *The Technologist.*] *Pharm. Journ.* II. vii. (1866) 479.

Pimento. [From *The Technologist.*] *Pharm. Journ.* II. vii. (1866) 616–618.

Peruvian Cinchonas. [Extract.] *Pharm. Journ.* II. viii. (1866) 80–81.

Erythroxylon Coca. [*London Medical Press.*] *Pharm. Journ.* II. viii. (1866) 299.

Beet Sugar and Cane Sugar. [From *Travers's Circular.*] *Pharm. Journ.* II. viii. (1866) 300–301.

Than-mo, a vermifuge remedy of the Burmese. [Extract.] *Pharm. Journ.* II. viii. (1866) 354.

Application of Turnips in dyeing. [Extract.] *Pharm. Journ.* II. viii. (1866) 355.

On the febrifuge properties of the leaves of Cinchona. [Extract.] *Pharm. Journ.* II. viii. (1866) 356–358.

Breve noticia sobre a collecção das madeiras do Brazil a presentada na
    Exposição Internacional de 1867.   Rio Janeiro, 1867.   8°.   *
Jurors' Report on the Vegetable Products in the Intercolonial Exhibi-
    tion of 1866.   [Melbourne, 1867.]   8°.
Beetroot Sugar.   [From *The Grocer.*]   *Pharm. Journ.* II. viii. (1867)
    733–734.
Australian Spinach.   [Extract.]   *Pharm. Journ.* II. viii. (1867) 734.
Ginseng.   [Extract.]   *Pharm. Journ.* II. ix. (1867) 77.
Artificial culture of Truffles.   [Extract from *The Grocer.*]   *Pharm.
    Journ.* II. ix. (1867) 77–78.
Exposition universelle de 1867.   Catalogue raisonné des collections
    exposées par l'administration des forêts.   Paris, 1868.   8°.
Japanese Tea.   [*The Grocer.*]   *Pharm. Journ.* II. ix. (1868) 343.
Truffle Hunt at Cannes.   [*Gardeners' Chronicle*].   *Pharm. Journ.* II.
    x. (1868) 29–30.
Aloes.   [*Med. Times and Gazette.*]   *Pharm. Journ.* II. x. (1868)
    106–110.
The introduction of Cinchona to the island of Jamaica.   *Pharm.
    Journ.* II. x. (1868) 166–168.
Proceedings of Conferences held by the India Committee of the
    Society [of Arts], on subjects connected with the Arts, Manufac-
    tures, and Commerce of India.   London, 1869.   8°.

Notes on Tea, Fibres, and Cotton.

Opium in China.   [*Times.*]   *Pharm. Journ.* II. xi. (1869) 304–305.
Edible Fungi.   [*Gardeners' Chronicle.*]   *Pharm. Journ.* II. x. (1869)
    431–433.
Kurze Anleitung zur Zucht und Pflege des Maulbeerbaums.   Leipzig,
    1870.   4°.
Resumen de los trabajos verificados por la comision de la flora forestal
    española durante los años de 1867 y 1868.   Madrid, 1870.   4°.
Brick Tea.   [*Food Journal.*]   *Pharm. Journ.* II. xi. (1870) 658.
Cultivation of Ipecacuanha in India.   [*Medical Press.*]   *Pharm. Journ.*
    III. i. (1870) 5.
Ashycrown cinchona in Venezuela.   [From *Vargasia.*]   *Pharm Journ.*
    III. i. (1870) 66.
Mushrooms.   [Edible species.]   *Pharm. Journ.* III. i. (1870) 88–89.
Silk and Sunflowers in Mauritius.   [Extract.]   *Pharm. Journ.* III. i.
    (1870) 130.
Indian Drugs.   [Speeches in Parliament.]   *Pharm. Journ.* III. i.
    (1870) 137.
Poppy culture in North America.   *Pharm. Journ.* III. i. (1870) 148.

Introduction of the Ipecacuanha plant into India. [*Allen's Indian Mail.*] *Pharm. Journ.* III. i. (1870) 170.

Ancient use of odoriferous plants. [Extract.] *Pharm. Journ.* III. i. (1870) 188.

Gingilie oil. [*Jaffna News.*] *Pharm. Journ.* III. i. (1870) 226.

Cinchona cultivation in India. *Pharm. Journ.* III. i. (1870) 325.

Java Cinchona bark. [Extract.] *Pharm. Journ.* III. i. (1870) 342-343.

East Indian Cinchona bark [sale of, grown at Darjeeling]. *Pharm. Journ.* III. i. (1870) 326.

The Mullein plant. [Verbascum Thapsus. From *N.Y. Druggists' Circular.*] *Pharm. Journ.* III. i. (1870) 365.

Monkey nuts. [Arachis hypogaea. From *Nature.*] *Pharm. Journ.* III. i. (1870) 488.

Report of the Commissioners appointed to enquire into the preparation of the Phormium tenax or New Zealand Flax. Wellington, 1871. fol. *

Catalogue of the samples of Fibres and manufactured articles prepared from the Phormium tenax; exhibited by the Flax Commissioners, in the Colonial Museum, Wellington, August, 1871. Wellington, 1871. 8°.

Catalogue of the Victorian exhibits to the Sydney Intercolonial Exhibition of 1870. Melbourne, 1871. 8°.

> Chemical and pharmaceutical products, pp. 52-54. Raw products and forest industries, pp. 54-56. Fibrous substances, pp. 57-63.

Le malettie delle piante, natura delle medisme, cause che le producono e processo di trattamento a prevenirle ed a curare; opera compilata sulle più accreditate antiche e moderne e specialmente sopra quelle dei professori Re e Moretti. Vol. i. Milano, 1871. 32°.

> Enciclopedia agraria popolare. Serie 1. vol. ii.

Botany in Medical Schools. [*Gardeners' Chron.*] *Pharm. Journ.* III. i. (1871) 546.

Vegetable wax in Japan. [From *Nature.*] *Pharm Journ.* III. i. (1871) 568.

Chinese native Opium. *Pharm. Journ.* III. i. (1871) 604-605.

The Guava. [*Druggists' Circular.*] *Pharm. Journ.* III. i. (1871) 605-606.

Californian Acorns. [*Nature.*] *Pharm. Journ.* III. i. (1871) 686.

Geranium dissectum. [*Nature.*] *Pharm. Journ.* III. i. (1871) 686.

Bhang- and Opium-eating in India. [*Medical Times and Gazette.*] *Pharm. Journ.* III. i. (1871) 706-707.

Kameela. [*Manufacturer and Builder.*] *Pharm. Journ.* III. i. (1871) 707–708.

The Bunya-Bunya (Araucaria Bidwilli). [*Nature.*] *Pharm. Journ.* III. i. (1871) 770.

Tuba roots. [*Gardeners' Chronicle.*] *Pharm. Journ.* III. i. (1871) 790.

Sumbulus moschatus. [Extract.] *Pharm. Journ.* III. i. (1871) 807.

Chinese products. [Extracts from Consular Reports.] *Pharm. Journ.* III. i. (1871) 807–808.

The Candleberry tree. (Aleurites triloba.) [*Gardeners' Chronicle.*] *Pharm. Journ.* III. i. (1871) 848.

Galuncha. [Tinospora cordifolia, *Miers.* From *The Gardeners' Chronicle.*] *Pharm. Journ.* III. i. (1871) 848.

Turmeric. [From *The Gardeners' Chronicle.*] *Pharm. Journ.* III. i. (1871) 868.

The Ink plant. [Coriaria myrtifolia. Extract from *Nature.*] *Pharm. Journ.* III. i. (1871) 928.

American Sumac. [*Scientific American.*] *Pharm. Journ.* III. i. (1871) 971–972.

Poppy farming in Queensland. [Extract.] *Pharm. Journ.* III. i. (1871) 972.

Sunflower-seed oil. [*Journ. Applied Science.*] *Pharm. Journ.* III. ii. (1871) 106.

Cultivation of Ipecacuanha in India. *Pharm. Journ.* III. ii. (1871) 227.

More about Condurango. *Pharm. Journ.* III. ii. (1871) 272.

Cinchona cultivation in Bengal. [Report.] *Pharm. Journ.* III. ii. (1871) 304–308.

Cinchona culture in the East. *Pharm. Journ.* III. ii. (1871) 404–405.

The Poppy crop in Behar and Benares. [Its partial failure through blight.] *Pharm. Journ.* III. ii. (1871) 429.

Paper manufacture in Japan. *Pharm. Journ.* III. ii. (1871) 526–528.

Progetto per la coltivazione ed industria dei prodotti della Asclepiade di Siria. Cremona, 1872. 8°.

Condurango. *Pharm. Journ.* III. ii. (1872) 665.

Cinchona cultivation in Java. *Pharm. Journ.* III. ii. (1872) 684.

Cinchona and Ipecacuanha in India. *Pharm. Journ.* III. ii. (1872) 689.

The Orange. [*Good Health.*] *Pharm. Journ.* III. ii. (1872) 768–769.

Olibanum. [Extract.] *Pharm. Journ.* III. ii. (1872) 867.

Poppy culture in Australia. *Pharm. Journ.* III. ii. (1872) 1028.

The manufacture of Attar of Roses in Turkey. *Pharm. Journ.* III. ii. (1872) 1051–1052.

The manufacture of Olive oil in California. [*Scientific American.*] *Pharm. Journ.* III. iii. (1872) 387–388.

The Chinese Materia medica. [Extract.] *Pharm. Journ.* III. iii. (1872) 501–502.

Persian Opium. [Extract.] *Pharm. Journ.* III. iii. (1872) 31.

Indian Opium. *Pharm. Journ.* III. iii. (1872) 248.

The collapse of Cundurango. *Pharm. Journ.* III. iii. (1872) 248.

The Lavender country. [*Journ. Applied Science.*] *Pharm. Journ.* III. iii. (1872) 325–326.

Pharmaceutical utility of botanical gardens. *Pharm. Journ.* III. iii. (1872) 328.

Ipecacuanha cultivation in India. *Pharm. Journ.* III. iii. (1872) 328.

Papers regarding the Tea industry in Bengal, *etc.* Calcutta, 1873. 8°.

Indian Opium [note on sale of]. *Pharm. Journ.* III. iii. (1873) 523.

Betel nut chewing. [Extract.] *Pharm. Journ.* III. iii. (1873) 547–548.

Opium trade of India. [*Times.*] *Pharm. Journ.* III. iii. (1873) 704.

Poppy culture in China. *Pharm. Journ.* III. iii. (1873) 890.

The Olive oil trade. *Pharm. Journ.* III. iii. (1873) 955.

Uses of the genus Vinca. *Pharm. Journ.* III. iii. (1873) 961.

Cultivation of Cinchona at Darjeeling. [*Times.*] *Pharm. Journ.* III. iv. (1873) 6.

Projected herbarium of medicinal plants [at the Pharmaceutical Society]. *Pharm. Journ.* III. iv. (1873) 8.

Indian Opium. [Note from *Journ. Agri. Hort. Soc. India.*] *Pharm. Journ.* III. iv. (1873) 32.

Rheea or China Grass. [Note on an advertisement in *The Times.*] *Pharm. Journ.* III. iv. (1873) 32.

Report on the Olive oil trade. *Pharm. Journ.* III. iv. (1873) 51.

Poisonous nuts. [Denna nuts,＝Arachis hypogaea.] *Pharm. Journ.* III. iv. (1873) 70.

The sanitary value of flowers. [Extract.] *Pharm. Journ.* III. iv. (1873) 64–65.

Camphor. [Extract.] *Pharm. Journ.* III. iv. (1873) 145.

The Lavender fields of Hertfordshire. [*Chambers's Journal.*] *Pharm. Journ.* III. iv. (1873) 165–166.

New Zealand native drugs. [Extract.] *Pharm. Journ.* III. iv. (1873) 186.

Olive oil from Tunis. *Pharm. Journ.* III. iv. (1873) 204–205.

Cocoa [The trade in]. *Pharm. Journ.* III. iv. (1873) 245–246.

Chinese Opium trade. *Pharm. Journ.* III. iv. (1873) 285–286.

Mesquite Gum. [From Prosopis glandulosa, *Torr.*] *Pharm. Journ.* III. iv. (1873) 286.

Flowering of Jalap plants in France. *Pharm. Journ.* III. iv. (1873) 303.

Cinchona cultivation. *Pharm. Journ.* III. iv. (1873) 323.

Flowering of the Scammony plant. *Pharm. Journ.* III. iv. (1873) 323.

Cinchona cultivation in Java. *Pharm. Journ.* III. iv. (1873) 341.

Patchouli. [*Journ. Applied Science.*] *Pharm. Journ.* III. iv. (1873) 362.

Olive oil from Crete. *Pharm. Journ.* III. iv. (1873) 385–386.

Trade in drugs, etc., with British India. *Pharm. Journ.* III. iv. (1873) 407–408.

The Eucalyptus Globulus as a disease-destroying tree. *Pharm. Journ.* III. iv. (1873) 494–495.

Palermo Manna. *Pharm. Journ.* III. iv. (1873) 496.

Opium trade in India. *Pharm. Journ.* III. iv. (1873) 498.

Cultivation of Vanilla in the island of Réunion. *Pharm. Journ.* III. iv. (1873) 517–518.

The Palm oil trade in Old Calabar. *Pharm. Journ.* III. iv. (1873) 531–532.

Las plantas industriales, tradado curiosa del cultivo y aprovechamiento de las plantas testiles, oleaginosas, tintorias y otras que son objeto de la industria. Madrid, 1874. 8°.

Saffron growing in France. [*Journ. Soc. Arts.*] *Pharm. Journ.* III. iv. (1874) 551.

Chinese Camphor. [Extract from a Report.] *Pharm. Journ.* III. iv. (1874) 589.

Notes on the Chinese Opium trade. *Pharm. Journ.* III. iv. (1874) 675.

Castor-oil seed. Ricinus communis. *Pharm. Journ.* III. iv. (1874) 676.

Ginger. Zingiber officinale. [Extract.] *Pharm. Journ.* III. iv. (1874) 676.

Preparation of Indian Hemp. [Extract.] *Pharm. Journ.* III. iv. (1874) 696–697.

Safflower. Carthamus tinctorius. *Pharm. Journ.* III. iv. (1874) 714.

The Eucalyptus plantations of Algeria. [Extract.] *Pharm. Journ.* III. iv. (1874) 731.

Poppy seed. Papaver somniferum. [Extract.] *Pharm. Journ.* III. iv. (1874) 731.

The coppicing of Cinchonas. [Extract.] *Pharm. Journ.* III. iv. (1874) 731–732.

Jaborandi, a new drug from Brazil. Pilocarpus pinnatus, *Linn.* [*Repert. de Pharm.*] *Pharm. Journ.* III. iv. (1874) 850.

Ailanthus glandulosa as a remedy for dysentery. [*Repert. de Pharm.*] *Pharm. Journ.* III. iv. (1874) 890.

The growth of Cinchona. [Rate of growth.] *Pharm. Journ.* III. iv. (1874) 894.

Cultivation of the Tea plant in Anjou. *Pharm. Journ.* III. iv. (1874) 1019.

The Vanilla. [*Gardeners' Chron.*] *Pharm. Journ.* III. v. (1874) 24–25.

Ceylon products. [Abstract.] *Pharm. Journ.* III. v. (1874) 47–48.

Japanese vegetable wax. [*Journ. Soc. Arts.*] *Pharm. Journ.* III. v. (1874) 166.

Herb cultivation at Mitcham. [*The Garden.*] *Pharm. Journ.* III. v. (1874) 182.

Report on the Government Cinchona plantations in Java. For the second quarter of the year 1874. *Pharm. Journ.* III. v. (1874) 282.

Cinchona cultivation in St. Helena. *Pharm. Journ.* III. v. (1874) 305–306.

Larch bark. *Pharm. Journ.* III. v. (1874) 328.

Olibanum. *Pharm. Journ.* III. v. (1874) 350.

The trade in drugs, etc., with Turkey. *Pharm. Journ.* III. v. (1874) 388.

Boldo. [Peumus Boldus.] *Pharm. Journ.* III. v. (1874) 405–406.

Japanese vegetable wax. [*Journ. Soc. Arts.*] *Pharm. Journ.* III. v. (1874) 425.

Acclimatization of medical plants [in the south of France]. *Pharm. Journ.* III. v. (1874) 448.

The wild Vanilla plant [Liatris odoratissima. From *The Garden.*] *Pharm. Journ.* III. v. (1874) 489–490.

Copal. [*Gardeners' Chronicle.*] *Pharm. Journ.* III. v. (1874) 490.

Opium in China. *Pharm. Journ.* III. v. (1874) 509–510.

The Sumbul plant. [Euryangium Sumbul, *Kauff.* Extract.] *Pharm. Journ.* III. v. (1874) 510.

Matico. [*Gardeners' Chronicle.*] *Pharm. Journ.* III. v. (1874) 523–524.

De la culture de la betterave au point de vue du rendement. Valenciennes, 1875. 12°.

Der praktische Forstwirth fuer die Schweiz. Lenzburg, 1875. 8°.

The Vegetable Kingdom; or, the produce of the earth and its uses. London, 1875. 4°.

The Jalap plant. (Exogonium purga.) [*The Garden.*] *Pharm. Journ.* III. v. (1875) 547–548.

The new drug, Jaborandi. *Pharm. Journ.* III. v. (1875) 569. *See also* p. 574.

The cultivation of Opium and Cinchonas in India. *Pharm. Journ.* III. v. (1875) 663–664.

Notes on some United States' drugs. [Extract.] *Pharm. Journ.* III. v. (1875) 704.

The botanical source of medicinal Rhubarb. [*Gardeners' Chronicle.*] *Pharm. Journ.* III. v. (1875) 784–785.

Cinchona in Madeira. *Pharm. Journ.* III. v. (1875) 835.

The true Gamboge plant. [Garcinia Hanburyi.] *Pharm. Journ.* III. (1875) 972.

Serronia Jaborandi. [Abstract.] *Pharm. Journ.* III. v. (1875) 1034–1035.

A new Mexican drug. [Damiana.] *Pharm. Journ.* III. vi. (1875) 24.

Flowering of the Euryangium Sumbul, *Kauffmann,* in England. [*Gardeners' Chronicle.*] *Pharm. Journ.* III. vi. (1875) 43–44.

The China Opium trade. *Pharm. Journ.* III. vi. (1875) 47–48.

The India Museum at South Kensington. [Native produce.] *Pharm. Journ.* III. vi. (1875) 61–62, 101–102, 181–182, 241–242, 381–382, 422–423.

Cultivation of Cinchonas in the Isle of Bourbon. [*Gardeners' Chron.*] *Pharm. Journ.* III. vi. (1875) 66–67.

Cinchona cultivation in Ceylon. [*Gardeners' Chron.*] *Pharm. Journ.* III. vi. (1875) 86.

Officinal Rhubarbs. [*Journ. Bot.*] *Pharm. Journ.* III. vi. (1875) 146.

Japanese edibles. [*Gardeners' Chron.*] *Pharm. Journ.* III. vi. (1875) 202, 221–222.

Cinchona bark from Santa Martha. [Extract.] *Pharm. Journ.* III. vi. (1875) 284.

Tahiti products. [Short note.] *Pharm. Journ.* III. vi. (1875) 284.

Coto bark. A reputed new remedy for Diarrhoea, Rheumatism and Gout. [*Archiv der Pharm.*] *Pharm. Journ.* III. vi. (1875) 301–303.

The indigenous Materia medica, etc., of Victoria. *Pharm. Journ.* III. vi. (1875) 368.

Tayuya. *Pharm. Journ.* III. vi. (1875) 401.

Damiana. [Extract.] *Pharm. Journ.* III. vi. (1875) 423.

Cinchona cultivation in private plantations [in India]. *Pharm. Journ.* III. vi. (1875) 447–448.

The competition between Indian and native-grown Opium in China. [Extracted from Consular reports.] *Pharm. Journ.* III. vi. (1875) 465–467.

Twelve prize essays on Sugar culture. Georgetown, 1876. [8°. ?] *

Mémoire sur le Silphium cyrenaïcum. Paris, 1876. 4°.

Musée colonial ou Pavillon près de Harlem. Notice sur les collections du musée pour servir de guide aux visiteurs. Harlem, 1876. 8°.

Productions végétales, pp. 7-35.

Note sur la culture des tabacs. Exposition permanente des colonies. Commission mixte des tabacs. Paris, 1876. 8°.

Philadelphia International Exhibition, 1876. Mexican section. Special catalogue and explanatory notes. Philadelphia, 1876. 8°.

Medicine, pp. 61-71. Agriculture, pp. 91-120.

Skogsvaennen. Stockholm, 1876. 8°.

"Green Tea." [*Shanghai Commonwealth.*] *Pharm. Journ.* III. vi. (1876) 746–747.

The name Jaborandi. *Pharm. Journ.* III. vi. (1876) 750.

Borneo Camphor. [Extract.] *Pharm. Journ.* III. vi. (1876) 772.

The earth Almond. [Cyperus esculentus, *L.*] *Pharm. Journ.* III. vi. (1876) 772. *See also* p. 748.

The Vanilla culture. *Pharm. Journ.* III. vi. (1876) 772.

Acorn Coffee. *Pharm. Journ.* III. vi. (1876) 872.

African essences. *Pharm. Journ.* III. vi. (1876) 872.

Persian opium. [Extracted from Consular Report.] *Pharm. Journ.* III. vi. (1876) 890.

The Eucalyptus Globulus; its hygienic and medical properties. [Abstract.] *Pharm. Journ.* III. vi. (1876) 912.

Persian opium. *Pharm. Journ.* III. vi. (1876) 950.

The Cultivation of Jaborandi and Eucalyptus. *Pharm. Journ.* III. vii. (1876) 10.

Resources of the province of Shantung. *Pharm. Journ.* III. vii. (1876) 26–27.

Aleppo Scammony, *etc.* [Extract from a Consular report.] *Pharm. Journ.* III. vii. (1876) 102.

Brazilian Tapioca. [Extract from a Consular Report.] *Pharm. Journ.* III. vii. (1876) 102.

The effect of age on Rhamnus Frangula bark. *Pharm. Journ.* III. vii. (1876) 102.

The Sunflower and its uses. [Extract.] *Pharm. Journ.* III. vii. (1876) 117.

Ava, or Kava-Kava. *Pharm. Journ.* III. vii. (1876) 149–150.

Acclimatization of plants. *Pharm. Journ.* III. vii. (1876) 160.

The Eucalyptus Globulus in Italy. *Pharm. Journ.* III. vii. (1876) 160.

Xanthium spinosum. *Pharm. Journ.* III. vii. (1876) 335.

Erythroxylon Coca. *Pharm. Journ.* III. vii. (1876) 335–336.

An extraordinary Cinchona bark. [Cinchona Ledgeriana, *Trimen.*] *Pharm. Journ.* III. vii. (1876) 360.

The end of the Cinchona experiment in St. Helena. *Pharm. Journ.* III. vii. (1876) 360.

Ipecacuanha and Vanilla cultivation in India. [Extract.] *Pharm. Journ.* III. vii. (1876) 433.

Report on Oil seeds in the India Museum. *Pharm. Journ.* III. vii. (1876) 468. *See also* COOKE.

The Copaiba tree. *Pharm. Journ.* III. vii. (1876) 516.

Caña de azúcar. Plantacion y cultivo de la caña de azúcar. Valencia, 1877. 8°.

Liberian Coffee. [*Journ. Soc. Arts.*] *Pharm. Journ.* III. vii. (1877) 574.

Cultivation of the Castor-oil bean. [*Chem. and Druggist.*] *Pharm. Journ.* III. vii. (1877) 592–593.

Indian Hemp and its active principle. [Extract.] *Pharm. Journ.* III. vii. (1877) 836.

Cinchona cultivation in Java. *Pharm. Journ.* III. vii. (1877) 861.

Pitury, an Australian rival to Coca. [Extract.] *Pharm. Journ.* III. vii. (1877) 878.

The collection of Masterwort and other roots on the continent. *Pharm. Journ.* III. vii. (1877) 986.

The sanitary influence of the Eucalyptus. [Extract.] *Pharm. Journ.* III. vii. (1877) 986.

Timbo, its properties and composition. [Extract.] *Pharm. Journ.* III. vii. (1877) 1020.

Mushroom culture in Japan. [*Journ. Soc. Arts.*] *Pharm. Journ.* III. vii. (1877) 1026.

Oranges and Lemons. *Pharm. Journ.* III. viii. (1877) 6.

Cultivation of Jalap in Manilla. *Pharm. Journ.* III. viii. (1877) 6.

Acclimatization experiments with medicinal plants. [Extract.] *Pharm. Journ.* III. viii. (1877) 30.

The manufacture of Palm oil in Western Africa. *Pharm. Journ.* III. viii. (1877) 68–69.

French Lactucarium. [Extract.] *Pharm. Journ.* III. viii. (1877) 202.

The carrageen crop. [*Scientific American.*] *Pharm. Journ.* III. viii. (1877) 304.

Historical notes on Opium. *Pharm. Journ.* viii. (1877) 347–348.

Cultivation of the Ipecacuanha plant. *Pharm. Journ.* III. viii. (1877) 366.

The commercial history of a Cinchona plantation. *Pharm. Journ.* III. viii. (1877) 409–410.

Liberian Coffee in Ceylon; the history of the introduction and progress of the cultivation up to April, 1878, *etc.* Colombo, 1878. 12°.

Catalogue of specimens of Indian foreign produce sent to the Paris Exhibition. Calcutta, 1878. 8°. *

Catalogue des végétaux ligneux indigènes et exotiques existant sur le domaine forestier des Barres-Vilmorin. Paris, 1878. 8°.

Statistique forestière. Paris, [1878–79.] 2 vols. 4°. Atlas fol.

The Paris Exhibition. Crude materials applicable in Medicine and Pharmacy. *Pharm. Journ.* III. ix. (1878) 21–23, 41–46, 82–86, 101–104. Essential oils, Perfumery, *etc. Ib.* 282–286. Crude materials used or applicable in the Arts. *Ib.* 301–307.

The botanical source of Tobacco. *Pharm. Journ.* III. viii. (1878) 710.

The culture of Cinchonas in Java. [Abstract.] *Pharm. Journ.* III. viii. (1878) 774.

Collection of Canada Balsam. [*Trans. Amer. Pharm. Assoc.*] *Pharm. Journ.* III. viii. (1878) 813.

Notes on the Olive [cultivation at San Remo]. *Pharm. Journ.* III. viii. (1878) 814.

Mustard. [*Hardwicke's Science Gossip.*] *Pharm. Journ.* III. viii. (1878) 852–853.

Note on a new Rhubarb. [Rheum hybridum, var. Colinianum, *Baill.*] *Pharm. Journ.* III. viii. (1878) 856.

Commercial Jaborandi. *Pharm. Journ.* III. viii. (1878) 892.

Opium production in Africa. [*Leeds Mercury.*] *Pharm. Journ.* III. viii. (1878) 1007–1008.

Caladium seguinum, or Dumb Cane. [Extract.] *Pharm. Journ.* III. viii. (1878) 1008.

The botanical source of Araroba. [*Repert. de Pharm.*] *Pharm. Journ.* III. viii. (1878) 1048.

Silphium laciniatum. (Compass Plant, Resin Weed.) [Extract.] *Pharm. Journ.* III. ix. (1878) 29–30.

Berberis Aquifolium. [Extract.] *Pharm. Journ.* III. ix. (1878) 68.

The Para and Ceara Rubbers, and Balsam of Copaiba trees. [*Journ. Soc. Arts.*] *Pharm. Journ.* III. ix. (1878) 86–89.

Production of Opium in China. *Pharm. Journ.* III. ix. (1878) 246.

Ballata, a substitute for Gutta Percha. [*Boston Journ. Commerce.*] *Pharm. Journ.* III. ix. (1878) 412.

Cultivation of the Cinchona in the United States. *Pharm. Journ.* III. viii. (1878) 572.

Der Schweizer Kraeutersammler. Ausfuehrliche Beschreibung aller in der Schweiz und den angrenzenden Laendern, auf den Bergen und in den Thaelen wildwachsenden Pflanzen und Kraeuter, nebst genauer Angabe ihres Gebrauches, Nutzens, ihrer Anwendung und Wirkung, ihres Anbaues, ihres Einsammlung, Aufbewahrung und Verwerthung. Neu-Ulm, 1879. 8°.

Catechismo forestale, ossia manuale popolare di selvicultura. Bergamo, 1879. 8°.

Della coltivazione de' gelsi ed alleramento dei bachi da seta; studii e suggerimenti. Milano, 1879. 16°.

The production of oil of Rosemary in Dalmatia. [Extract.] *Pharm. Journ.* III. ix. (1879) 618.

Orange flowers and Oranges from the Southern States. [Extract.] *Pharm. Journ.* III. ix. (1879) 915–916.

Argan oil. [Extract from *Gardeners' Chron.*] *Pharm. Journ.* III. x. (1879) 127–128.

Cultivation of Perfume plants in South Australia. [*Journ. Soc. Arts.*] *Pharm. Journ.* III. x. (1879) 185–186.

Fucus vesiculosus. [*Lancet.*] *Pharm. Journ.* III. x. (1879) 343.

Rose-farming as a Colonial industry. [*Colonies and India.*] *Pharm. Journ.* III. x. (1879) 469–470.

Ministro di agricoltura, industria e commercio. Direzione dell' agricoltura. Annali di agricoltura 1879. Esperienze di coltivazione di tabachi eseguite dalle stazione agrarie. Roma, 1880. 8°.

Die forstliche Verhaeltnisse Wuerttemburgs, *etc.* Stuttgart, 1880. 8°.

The cultivation of the Sorghum plant, and its manufacture into sugar in the United States. *Journ. Soc. Arts,* xxviii. (1880) 469–470.

Notes on useful plants. [Extracts from the Report on Kew Gardens, for 1879.] *Journ. Soc. Arts,* xxix. (1880) 88–91.

Indian drugs. *Pharm. Journ.* III. x. (1880) 777–778.

Wattles. [Acacia sp.] *Pharm. Journ.* III. x. (1880) 798.

Persian and native Opium in China. *Pharm. Journ.* III. x. (1880) 800.

The production of Indigo in Bengal. [*Journ. Soc. Arts.*] *Pharm. Journ.* III. x. (1880) 917–918.

Resin and Turpentine. [*Journ. Soc. Arts.*] *Pharm. Journ.* III. xi. (1880) 26.

The Cork tree. [Extract.] *Pharm. Journ.* III. xi. (1880) 134–136.

Experimental cultivation in Jamaica. *Pharm. Journ.* III. xi. (1880) 198.

Acclimatization experiments in Jamaica. *Pharm. Journ.* III. xi. (1880) 277–278.

Cinchona cultivation in Bengal. [Extract from report.] *Pharm. Journ.* III. xi. (1880) 334–335.

Ground-nuts. [*Gardeners' Chronicle.*] *Pharm. Journ.* III. xi. (1880) 494.

Cassia lignea. [Extract from Consular Report.] *Pharm. Journ.* III. xi. (1880) 498.

The uses of Henna in Algeria. [Abstract.] *Pharm. Journ.* III. xi. (1880) 515.

Cinchona bark market. *Pharm. Journ.* III. xi. (1880) 784.

Des marchès de betteraves au double point de vue de la culture et de la fabrication du sucre. Arras, 1881. 8°.

Il caffè di girasole: analisi chemiche, consigli agronomici, ecc. Padova, 1881. 8°.

Coca. (Erythroxylon Coca.) *Journ. Soc. Arts*, xxix. (1881) 472.

The cultivation of wheat-straw, and the manufacture of straw goods in Italy. *Journ. Soc. Arts*, xxix. (1881) 630–631.

Orange culture in Syria. *Journ. Soc. Arts*, xxix. (1881) 787.

The commerce of Gum Arabic at Trieste. [Extract.] *Pharm. Journ.* III. xi. (1881) 822.

Cinchona bark as an article of the official Materia medica. *Pharm. Journ.* III. xi. (1881) 903–904.

A Malagasy Materia medica. *Pharm. Journ.* III. xi. (1881) 853–855.

Cinchona cultivation in Ceylon. [Extract from a report.] *Pharm. Journ.* III. xii. (1881) 52.

Liquorice root. [From *The Oil and Drug News.*] *Pharm. Journ.* III. xii. (1881) 264.

ADDENDA.

ADAMS (—.).

    Cultivation of Cinchona in the United States. [Extract from a Consular report.] *Pharm. Journ.* III. xi. (1881) 1025.

ADELER (Christian Lente), & Jakob IVERSEN.

    Kurzgefaszte Anleitung zum Hopfenbau. Zunaechst als Wegweiser fuer die Mitglieder des Nord-Angler Hopfen-Vereins aus den neuesten Schriften ueber Hopfenbau zusammentragen. Flensburg, 1824. 8°. [*Anon.*]

AEGIDIUS (Peter Christian).

    Nogle Bemaerkninger over Kommen- og Rapsaed-Avlen. Kjoebenhavn, 1825. 8°.

ALLUAUD (—.), *aîné*.

    Mémoires sur le reboisement et la conservation des bois et forêts de la France. Limoges, 1845. 8°.

ANCELIN (Constant François).

    Réflexions agricoles et manufacturières sur l'industrie linière. Lille, 1856–7. 8°. (Three pamphlets.)

ARBOIS DE JUBAINVILLE (Alexandre d').

    Observation sur le systéme d'élagage de Courval et Des Cars. Paris, 1870. 16°.

    Note sur l'élagage des arbres forestiers. Paris, 1872. 18°.

ARNDT (Ernst Moritz).

    Ein Wort ueber die Pflegung und Erhaltung der Forsten und der Bauern in Sinne einer hoeheren und menschlichen Gesetzgebung. Schleswig, 1820. 8°.

ARNOLD (Edwin Lester).

    On the Indian Hills: or, Coffee planting in Southern India. London, 1881. 2 vols. 8°.

ARTUS (Wilhelm Friedrich Wilibald).

    Handatlas saemmtlicher medicinisch-pharmaceutischer Gewaechse. Ed. 6. by G. von Hayek. Jena, 1881. 8°.→

    The first edition is noticed on page 8.

BAGNERIS (Gustave). *See* p. 9.

BAKER (Edmund Gilbert).

An officinal Aloe from Madagascar. *Pharm. Journ.* III. xii. (1881) 43–44.

BÁLFOUR (—.), *Surgeon-Major.*

Manilla hemp. *Journ. Soc. Arts,* xxix. (1881) 661–662.

BAZELAIRE (Hippolyte de).

Traité du reboisement, ou manuel du planteur. Nancy, 1846. 12°. Ed. 2. Paris, 1864.

BEAUVISAGE (Georges Eugène).

Contribution à l'étude des origines botaniques de la Gutta-percha. Paris, 1881. 8°.

BECKER (Jens Fr.).

En kort Anviisning til Tobaks-Plantning. Viborg, 1809. 8°.

BEDÖ (Albert).

Die wirthschaftliche und commercielle Beschreibung der Koenigl.- Ungarischen Staatsforste. Budapest, 1878. 4°.

—— Description économique et commercial des forêts de l'ètat des Hongries, avec une tableau synoptique des forêts de l'état. (Trad. du texte original hongrois.) Budapest, 1878. 4°.

BELL (John).

A practical treatise on the culture of Sugar Cane, and distillation of rum . . . with some remarks on the cultivation of Cotton, *etc.* Calcutta, 1831. 8°.

BENNETT (John Whitchurch).

A treatise on the Coco-nut tree, and the many valuable properties possessed by that splendid palm, ascertained by personal observation. By a Fellow of the Linnaean and Horticultural Societies, many years resident in the island of Ceylon. With an interesting traditional account of its original discovery, by a prince of the interior of that island. London, 1831. 8°. [*Anon.*]

BERNARDIN (R. J.).

Classification de 350 matières tannantes. Gand, 1880. 8°.

This is a revised version of the work given on p. 19.

BERRY-RAYNAL (—.).

Mémoire sur un nouvel aménagement des bois et forêts, ou l'art d'augmenter les produits forestiers. Châlon-sur-Saône, 1858. 8°.

BIANCA (Giuseppe).

Il carrubo. Monografia storico-botanico-agraria. Firenze, 1881. 8°.

Bignone (Felix).

Note on the Corsican Moss of the Pharmacies. [Marine algae.] *Pharm. Journ.* III. xii. (1881) 258.

Bjerregaard (Hans).

En kort Anviisning til Traeavl eller til de almindeligste Frugt- og Skovtraeers Opelskning og Udplantning. Bestemt til Uddeling blandt Landalmuen i Danmark. Randers, 1828. 8°.

Bobierre (Adolphe).

Essai sur la culture de la canne à sucre par M. A. Reynoso, resumés critiques de MM. B., Girard et Barral. Paris, 1865. 8°.

Bojer (Wenzel).

Notice on the probable cause of the phenomenon manifested in the Sugar-Cane. [Blight.] *Trans. Royal Soc. Mauritius*, vol. i. (pt. ii.) (1845) 116–123.

Booth (John).

Feststellung der Anbauwuerdigkeit auslaendischer Waldbaeume. Referat, *etc.* Berlin, 1880. 8°.

Boucard (—.).

De la sylviculture dans le département de l'Indre. Chateauroux, 1865. 8°.

Bouché (Carl David). *See* p. 24.

Bourlier (Charles).

Guide pratique de la culture du lin en Algérie. Alger, 1863. 8°.

Bowman (Frederic Hungerford).

The structure of the Cotton fibre in its relation to technical applications. Manchester, 1881. 8°.

Brenot (L.).

Remarques sur deux variétés d'épicéa. Paris, 1878. 4°.

Briers (F.).

Aperçu sur l'élagage et la conduite des arbres forestiers et autres arbres destinés à l'industrie. Bruxelles, 1870. 8°.

Brown (John Ednie).

Timber Trees of South Australia. Adelaide, 1880. 8°.

A Practical Treatise on Tree Culture in South Australia. Adelaide, 1881. 8°.

Browne (D. J.).

The history, industry, and commerce of Flax. *Reports, Commissioners of Patents, Agriculture.* Washington, 1862. pp. 21–83.

BRETSCHNEIDER (E.).

Notes on some botanical questions connected with the export
trade of China. Peking, 1880. 8°.

On Chinese silkworm trees. Peking, 1881. 8°.

BROOKES (R.).

The natural history of Chocolate. Being a distinct and particular
account of the Cocao-tree, its growth and culture, and the
preparation, excellent properties, and medicinal vertues of its
fruit. Wherein the errors of those who have wrote upon this
subject are discover'd; the best way of making chocolate is
explain'd; and several uncommon medicines drawn from it, are
communicated . . . Translated from the last edition of the
French by a Physician [R. Brookes]. London, 1724. 8°.
Ed. 2. [Title slightly varied,] 1730.

> The first edition was anonymous.

BRUEEL (Fr.).

Gekroente Preisschrift . . . ueber die beste Art die Waelder
anzupflanzen, zu nutzen und im Stande zu erhalten. Ed. 3.
Copenhagen und Leipzig, 1799. 8°.

BRUEEL (Georg Wilhelm).

Bidrag til den practiske Forstvidenskab. Kjoebenhavn, 1802. 8°.

—— Abhandlungen fuer Freunde der praktischen Forstwissenschaft. Kopenhagen, 1802. 8°.

BURGESS (T. J. W.).

The beneficent and toxical effects of the various species of
Rhus. [*Canadian Journ. Med. Science.*] *Pharm. Journ.* III.
xi. (1881) 858–860.

BURGSDORFF (Friedrich August Ludwig von).

Anviisning till at opelske indenlandske og udenlanske Traeearter
i det Frie. Oversat og omarbeidet til Anvendelse for Dannemark og Holsteen af Mart. Gottl. Schaeffer. Forsynet med
Anmaerkninger af Erik Viborg. (1 Deel.) Kjoebenhavn,
1799. 8°.

BURKART (—.).

Sammlung der wichtigsten europaeischen Nutzhoelzer in charak-
teristischen Schnitten. Bruenn, 1880. 8°.

CALVET (A.).

Note sur la culture du bambou et ses usages industriels dans
la region des Pyrénées et dans le sud-ouest de la France.
Paris, 1878. 4°.

CARPENTIER (Ernest de).

Plantation des terrains crayeux de la Champagne et des marais du nord de la France ; resultats économiques et financiers ; voeux et réformes. Paris, 1881. 18°.

CARVER (Jonathan).

A treatise on the culture of the Tobacco plant . . . adapted to northern climates. London, 1779. 8°. [Ed. 2.] Dublin, 1779.

—— Afhandling om Dyrknings- og Behandlings-Maaden af Tobak-Planten passelig til de Nordlige Himmel-Egue oversat af de Engelske efter J. C., *etc.* Christiania, 1781. 8°.

CECH (C. O.).

Ueber den Ursprung der Hopfencultur. Muenchen, 1881. 8°.

CHAMBERLAYNE (John).

The manner of making of Coffee, Tea and Chocolate . . . with their vertues. Newly done out of French and Spanish. London, 1685. 18°.

> The tracts on Tea and Coffee are from the French of P. SYLVESTRE DUFOUR, and that on Chocolate from the Spanish of A. COLMENERO DE LEDESMA.

CHAMBRAY (Georges de).

Traité pratique des arbres résineux conifères à grandes dimensions, *etc.* Paris, 1845. 8°.

CHÉLUS (—de).

Histoire naturelle du cacao et du sucre, devisée en deux traités. [With additions by N. Mahudel.] Paris, 1719. 8°. Ed. 2. Amsterdam, 1720.

CHEVALLIER (J. B A), *see p. 36, is sometimes spelled* CHEVALIER.

CHRIST (Johann Ludwig).

Der neueste und beste deutsche Stellvertretter des indischen Caffe oder der Caffee von Erdmandeln ; zu Ersparung vieler Millionen Geldes fuer Deutschland und laengerer Gesundheit Tausender von Menschen. [Cyperus esculentus, *L.*] Ed. 2. Frankfurt-am-Mayn, 1801. 8°.

CLERGEAUD (P.).

De la culture du coton en Algérie. Paris, 1862. 8°.

COLLINS (James).

Cultivation of Caoutchouc trees in India. *Journ. Soc. Arts*, xxix. (1881) 363–364.

COLMENERO DE LEDESMA (Antonio).

Curioso tratado de la naturaleza y calidad del Chocolate, dividido in quatro puntes, *etc.* Madrid, 1631. 4°,

COLMENERO DE LEDESMA (Antonio) *continued :—*

—— A curious treatise of the nature and quality of Chocolate. Written in Spanish by A. C. . . . and put into English by Don Diego de Vades-forte [*pseud., i.e.* J. Wadsworth]. London, 1640. 4°.

—— Du chocolate. Discours curieux . . . traduit d'Espagnol en François . . . par R. Moreau, *etc.* Paris, 1643. 4°.

—— Chocolata Inda. Opusculum de qualitate et naturae chocolatae . . . nunc in Latinum translatum. [By J. G. Volckamer.] Norimbergae, 1644. 16°.

—— Chocolate ; or, an Indian drinke . . . written originally in Spanish . . . and faithfully rendered in the English by Capt. J. Wadsworth. London, 1652 [? 1655]. 12°.

> In the preface, the translator uses the pseudonym of Diego de Vadesforte.

—— Della cioccolata discorso . . . tradotto dalla lingua spagnuola nell' italiana, con aggiunta d'alcune annotationi da A. Vitrioli. Roma, 1667. 12°. [Ed. 2.] Bologna, 1694.

COMES (Orazio).

I funghi in rapporto all' economia domestica ed alle piante agrarie. Lezione ecc. raccolte e scritte dall' alunno, L. Savastano. Napoli, 1880. 8°.

COOK (D. M.).

The culture and manufacture of sugar from Sorghum. *Report, Commissioner of Patents, Agriculture.* Washington, 1862. pp. 311–319.

COTTA (Heinrich).

Principes fondamentaux de la science forestière. Ouvrage traduit par Jules Nouguier. Ed. 2. Paris, 1841. 8°.

For original German edition see p. 42.

COURVAL (Ernest Alexis, *Comte* de).

Taille et conduite des arbres forestiers et autres arbres de grandes dimensions, *etc.* Paris, 1861. 8°.

CRAIG (William).

The medicinal plants of England and Ireland. *Pharm. Journ.* III. xi. (1881) 863–866.

CRINON (J. L. F.), & Charles VASSEROT.

Le forestier practicien, ou guide des gardes champêtres, traitant de la conservation des semis, de l'aménagement, de l'éxploitation, etc., des foréts. (Manuel Roret.) Paris, 1852. 18°.

CROIZETTE (Desnoyers).

Notice sur les divers emplois du hétre. Paris, 1878. 4°.

Notice sur le débit et les emplois des principales espèces des pins. Paris, 1878. 4°.

Notice sur le gemmage du pin maritime. Paris, 1878. 4°.

CROSBY (Schuyler).

Cultivation of the Chestnut in Tuscany. [From Consular report.] *Journ. Soc. Arts*, xxix. (1881) 451–452.

CROY (*le comte* André Rodolphe Claude François Siméon, *styled* Raoul de).

Avenir forestier de la France, considéré dans ses rapports avec les essences résineuses. Paris, 1853. 12°.

CURIE (—.).

Des produits tirés du pin maritime. Essai théorique et pratique sur la fabrication des matières résineuses. Paris, 1874. 8°.

D. (N.).

The vertues of Coffee. Set forth in the works of The Lord Bacon, his Natural Hist., Mr. Parkinson, his Herbal, Sir George Sandys, his Travails, John Howel, Esq., his Epistle. London, 1663. 4°.

In the body of the work it is called Coffa.

DALGAS (Carl Fredrik Isaac).

Iagttagelser over Hampens Dyrkning, samlede paa en Reise i Tydskland, Helvetien og Frankrige i Aarene 1808, 1809 og 1810. Kjoebenhavn, 1812. 8°.

DARLINGTON (William).

Agricultural botany : an enumeration of useful plants and weeds. Philadelphia, 1847. 8°.

DÉCUGIS (Bernardin).

Les tourteaux des graines oleagineuses, et leurs applications théoriques et pratiques dans la culture, l'alimentation des animaux . . . l'économie domestique, *etc.* Toulon, 1876. 8°.

DELAMARRE (Louis Gervais).

Traité pratique de la culture des pins à grande dimensions, *etc.* Paris, 1826. 8°. Ed. 3. 1831.

Historique de la creation d'une richesse millionnaire par la culture des pins, ou application du traité pratique de cette culture publiée en 1826, *etc.* Paris, 1826. 8°. Supplement, 1827.

DELAMER (Eugene Sebastian).

Flax and Hemp, their culture and manipulation, *etc.* London, 1854. 8°.

DELCHEVALERIE (G.).

Le dattier, sa description, son histoire, *etc.*   Gand, 1873.   8°.

DEMONTZEY (P.).

Traité pratique du reboisement et du gazonnement des montagnes.
Ed. 2.   Paris, [1881].   8°.

>   This is the original work of which a German version is cited on
>   p. 50.

DES CARS (Amedée Joseph).

L'élagage des arbres ; traité pratique de l'art de deriger les arbres
forestiers, *etc.*   Paris, 1865.   32°.   Eds. 2, 3, & 4.   1865.   Ed.
5.   1866.

—— A treatise on pruning forest and ornamental trees.   Translated
from the 7th French edition, with an introduction by Charles
S. Sargent.   Boston, 1881.   8°.

DE VOS (C. de).

Beredeneerd woordenboek der voornaamste heesters en coniferen
in Nederland gekweekt.   Groningen, 1867.   8°.

DEY (Kanny Loll).

Notes on some Indian Drugs.   Wrightia antidysenterica, Psoralea
corylifolia, Symplocos racemosa.   *Pharm. Journ.* III. xii. (1881)
257–258.

DIDRICHSEN (Didr.).

Skovbog eller Anviisning til Skovs Opelskning isaer med Hensyn
til Smaaeskoves Anlaeg paa Boenderjorderne.   Kjoebenhavn,
1803.   8°.

DOWNES (—.).

The growth of Crocus sativus, the source of saffron, in Kashmir.
[*The Gardeners' Chronicle.*]   *Pharm. Journ.* III. xii. (1881) 9.

DU BREUIL (A.).

Arboriculture.   Paris, 1881.   8°.

DUFOUR (P. S.).   *See* SYLVESTRE DUFOUR.

ELOFFE (Arthur).

L'ortie, ses propriétés alimentaires, médicales, agricoles et industrielles.   Ed. 2.   Paris, 1869.   16°.

>   For Ed. 1. *see* page 60.   This author has also written under the
>   pseudonym of KROENISHFRANCK, *see* page 108.

ERNST (Adolph).

Las familias mas importantes del reino vegetal especialmente las
que son de interes en la medicina, la agricultura y industria o
que estan representadas en la flora de Venezuela.   Carácas,
1881.   8°.

ERNST (Adolph) *continued* :—

Memoria botanica sobre el embarbascar, ó sea la pesca por medio de plantas venenosas. Carácas, 1881. 8°.

EYKMAN (J. F.).

Illicium religiosum, *Sieb.*, its poisonous constituent, and essential and fixed oils. [Transl.] *Pharm. Journ.* III. xi. (1881) 1046–1050.

The botanical relations of Illicium religiosum, *Sieb.*, and Illicium anisatum, *Lour.* [Transl.] *Pharm. Journ.* III. xi. (1881) 1066–1068.

FAULKNER (Alexander).

A Dictionary of commercial terms, with their synonymes in various languages. Bombay, 1856. 8°.

FILLIAS (Achille).

Exposition universelle de Paris en 1878. Notice sur les foréts de l'Algérie ; leur étendu ; leurs essences ; leurs produits. Alger, 1878. 8°.

FISCHER (Theodor).

Die Dattelpalme, ihre geographische Verbreitung und cultur-historische Bedeutung. Gotha, 1881. 4°.

FLEISCHER (Elias).

Forsoeg til en Underviisning i det Danske og Norske Skov-Vaesen. Kjoebenhavn, 1779. 8°.

FLOR (Martin Richard).

Om Tobakavl ; og Humleavl af C. E. Wiinholt. Christiania, 1817. 8°.

FLUECKIGER (Friedrich A.), & Arthur MEYER.

Notes on the fruit of Strychnos Ignatii. *Pharm· Journ.* III. xii. (1881) 1–6.

FOREST (A.).

De la question du reboisement et nouvel examen des circonstances climatologiques et des faits économiques qui se rattachent à l'existence des foréts. Paris, 1852. 8°.

FRUEHLING (R.), & J. SCHULZ.

Anleitung zur Untersuchung der fuer die Zuckerindustrie in Betracht kommenden Rohmaterialen, Producte, Nebenproducte und Huelfssubstanten. Ed. 2. Braunschweig, 1881. 8°.

Ed. 1. is given on p. 68.

GALLAND (Antoine).

De l'origine et du progrez du café. Sur un manuscrit arabe de la Bibliothèque du Roy. Caen, 1699. 12°.

GALLOT (—.).

Notice sur le débit et les emplois du sapin, de l'épicéa, et du mélèze. Paris, 1878. 4°.

GALLOT (—.), & GAST (—.).

Notice sur le débit et les emplois du chêne rouvre et du chêne pédonculé. Paris, 1878. 4°.

GAUDRY (Albert).

Recherches scientifiques en Orient. Partie agricole. Paris, 1855. 8°.

> Chiefly connected with Cyprus.

GAUDRY (Louis).

Cours pratique d'arboriculture, etc. Paris, 1848. 12°. Ed. 2. 1849.

GAYFFIER (Eugène de).

Iconographie du reboisement et du gazonnement des montagnes. Descriptions, plans, et vues photographiques des grandes travaux dans les Alpes et les Pyrenées françaises. Paris, 1881. fol.

GERMAIN (Félix).

Rapport sur le reboisment, présenté au conseil général de la Drôme (août 1873). Valence, 1873. 8°.

GIHOUL (L.).

Culture forestière des arbres résineux conifères. Brèda, 1844. 8°.

GORGE-GRIMBLOT (A.).

Études sur la truffe. Paris, 1878. 4°.

GORKOM (Karel Wessel van).

De Oost-Indische cultures in betrekking tot handel en nijverheid. Amsterdam, 1881. 2 vols. 8°.

> The first volume is noticed on page 73; the portion relating to Cinchona will shortly be published in English.

GOSSON (Narcisse).

Quelques mots sur le coton et sur la colonisation. Le Havre, 1863. 8°.

GOURRIER (H.).   See HUBERT-GOURRIER (A.).

GRAETER (L.).

Ausfuehrliche Anleitung zur Kultur des Mohns um dieselbe auf das Doppelte des gewoehnlichen Ertrags zubringen, durch seine Benuetzing auf Opium. Nach dem Franzoesischen des A. Odeph . . . frei bearbeitet von einem Freunde der Landwirthschaft. Stuttgart, 1867. 8°.

GRANDVAUX (Emmanuel Louis).

Du reboisement des montagnes de France. Paris, 1846. 8°.

GREENISH (Henry George).

Contributions to the chemistry of Nerium odorum. *Pharm. Journ.* III. xi. (1881) 873–875.

GRIFFIN (—.).

The Kauri gum of New Zealand. [*Scientific American.*] *Pharm. Journ.* III. xi. (1881) 989.

GRIGOR (J.).

Arboriculture; or, a practical treatise on raising and managing forest trees, and on the profitable extension of the woods and forests of Great Britain. Edinburgh, 1881. 8°.

GURNAUD (Antoine).

Traité forestier pratique; manuel du proprietaire de bois. Paris, 1870. 8°.

HENDESS (H.).

Waaren-Lexikon fuer den Droguen-, Spezerei- und Farbwaarcn Handel, sowie der chemischen und technischen Praeparate fuer Apotheker. Berlin, 1881. 8°.

HEUZÉ (Gustave).

Les céréales, les produits farineux et leurs derivès à l'Exposition universelle internationale de 1878 à Paris. Paris, 1881. 8°.

HIDALGO (Tablada José de).

Tratado del cultivo del olivo en España, y modo de mejorarlo. Madrid, 1870.

—— Tradado del cultivo de los árboles frutales en España y modo de mejorales. Ed. 2. Madrid, 1871. 8°.

HOCHSTETTER (Ferdinand).

Grosses illustrirtes Kraeuterbuch. Ausfuehrliche Beschreibung der fuer Arzneikunde, Handel und Industrie verwendbaren Pflanzen und Mineralien, deren Fundort, deren Verwendung und der daraus gewonnenen Producte . . . Nach den neuesten Quellen herausgegeben von F. H., MARTIN und Anderen. Reutlingen, 1880. 8°.

HOCHSTETTER (W.).

Die Coniferen oder Nadelhoelzer, welche in Mitteleuropa winterhart sind. Stuttgart, 1881. 8°.

HOLMES (Edward Morell).

The varieties of Linseed in English commerce. *Pharm. Journ.* III. xii. (1881) 61–62, 137–140.

The cultivation of medicinal plants in Lincolnshire. *Pharm. Journ.* III. xii. (1881) 237–239.

Which kinds of Cinchona bark should be used in Pharmacy? *Pharm. Journ.* III. xii. (1881) 368–369.

HOLTZENDORFF (Julius von).
Der Flachs, sein Anbau, und seine Zubereitung in Ireland. Aus dem Englischen . . . uebertragen von J. von H. Leipzig, 1865. 8°.
Translated from CHARLEY, and WARD.
HORN (J. E.).
La crise cotonnière et les textiles indigénes. Paris, 1863. 8°.
HORNE (John).
A year in Fiji, or an inquiry into the botanical, agricultural, and economical resources of the colony. London, 1881. 8°.
HOWARD (John Eliot).
On [Cinchona] Red Bark. *Pharm. Journ.* III. xii. (1881) 350–354.→
HUBERT-GOURRIER (A.).
Traité de la culture de l'olivier et de la fabrication de l'huile d'olive. Toulon, 1881. 8°.

ICERY (Edmond). *See* p. 99.
IVERSEN (Jacob).
Om Rapsaedens Dyrkning i det Holsteenske, i Saerdeleshed i Hertugdoemmet Slesvig . . . oversat af det Tydske. Kjoebenhavn. 1803. 8°.

JACQUEMART (D. A.).
Bibliographie forestière française, ou catalogue chronologique des ouvrages français ou traduits en français et publiés depuis l'invention de l'imprimerie jusqu'à ce jour, sur la sylviculture, l'arboriculture forestière, et sur les matières qui s'y rattachent. Paris, 1852. 8°.
JAEGER (H.).
Deutsche Baeume und Waelder. Ed. 2. Leipzig, 1881. 8°.
JAEGER (Niels Knag).
Kort Anviisning om Maaden, sikkerst at forplante Traeer paa, saavel ved Saeden, som af unge Roenninger, grundet paa egue Forsoeg, *etc*. Bergen, 1778. 8°. [*Anon.*]
JARDIN (E.).
Le coton, son histoire, son habitat, son emploi et son importance chez les différents peuples, avec l'énumeration de ses succedanés. Genève, 1881. 12°.
JOHNSON (S. W.).
Tobacco. *Report of Chemist to Connecticut State Board of Agriculture,* 1873, pp. 383–424.

JOLIVET (—.).

Notice sur l'emploi du bois dans la fabrication du papier. Paris, 1877. 4°.

JOUYNE (Zéphirin).

Reboisement des montagnes. Reboisement, difficultés, causes des inondations et moyens de les prévenir. Digne, 1852. 8°.

KASTHOFER (Carl).

Bemerkungen ueber die Waelder und Alpen des Bernerischen Hochgebirges, *etc.* Ed. 2. Aarau, 1818. 8°.

——Guide dans les fôrets. Traduite par F. L. Monney. Vevey, 1830. 2 vols. 8°.

KING (George).

Cinchona cultivation in Bengal. [Extract from report.] *Journ. Soc. Arts*, xxix. (1881) 844–845.

KLOSE (M.).

Ein Wort ueber Lein- und Flachsbau, dessen Cultur, Bearbeitung und fernere Benutzung. Hirschberg, 1881. 8°.

KOLTZ (Jean Pierre Joseph).

Mémoire sur le boisement des terres incultes. Namur, 1866. 8°.

KROGH (Fr. Ferdinand von).

Kort Underviisning for Forstbedienterne udi det forste Slesvigske Jaegermester-District. Haderslev, 1800. 8°.

LANGGUARD (Alexander).

Japanese and Chinese aconite roots. [*Archiv der Pharm.*] *Pharm. Journ.* III. xi. (1881) 1021–1025, 1041–1045.

LARZILLIÈRE (—.).

Notice sur le débit des bois de feu, leur mode de vente, et les procédés, de carbonisation usités en France. Paris, 1878. 4°.

LE BRETON (F.).

Traité sur le propriétés et les effets du sucre avec le traité de la petite culture de la canne à sucre, *etc.* Paris, 1789. 12°.

LEON (John A.).

On Sugar cultivation in Louisiana, Cuba, etc., and the British Possessions. [2 parts. Part ii. is entitled 'The Sugar Question.'] London, 1848. 8°.

LESPINASSE (—.).

Hemp cultivation in Mexico. [From a Consular report.] *Journ. Soc. Arts*, xxix. (1881) 488–489.

LIOTARD (L.).

Memorandum on dyes of Indian growth and production.   Calcutta, 1881. fol.

Plantain tea. [Extract.]   *Journ. Soc. Arts*, xxix. (1881) 688–690.

LOCHNER (Michael Friedrich).

Schediasma de Parreira brava, novo Americano aliisque recentioribus calculi remediis, *etc.*   *Praes.* Lucae Schroek.   Ed. 2.   Norimbergae, 1719. 4°.

> The preface is dated 1712, which probably marks the date of the first edition.

De novis et exoticis Thee et Cafe succedaneis . . . dissertatio epistolica. Noribergae, 1717. 4°.

LOCK (Charles George Warnford).

Notes on Gums, Resins and Waxes. *Journ. Soc. Arts*, xxix. (1881) 657–661.

The Soy bean, a new feeding stuff. *Journ. Soc. Arts*, xxix. (1881) 734–735.

Esparto or Alfa. *Journ. Soc. Arts*, xxix. (1881) 787–788.

LOEBE (W.).

Die Futterkraeuter, Abbildung und Beschreibung aller in der Landwirthschaft vorkommenden und zu benutzenden Kraeuter. Ed. 3. Dresden, 1881. 8°.

> For earlier issues (which seem confused), see page 119.

McIVOR (R. W. E.).

An educational lecture on the Food of Plants in its relations to the exhaustion of lands . . . also, Select textile plants deserving extensive culture in the colony of Victoria, a lecture . . . by Baron Ferd. von Mueller. Ballarat, [1876.] 8°.

MAGRI (Domenico).

Virtv del Kafé, bevanda introdotta nvovamente nell' Italia . . . Seconda impressione con aggiunta del medesimo autore. Roma, 1671. 4°.

MAISCH (John M.).

Notes on the Xanthorrhoea resins. [*Amer. Journ. Pharm.*] *Pharm. Journ.* III. xi. (1881) 1005–1006. Also in *Journ. Soc. Arts*, xxix. (1881) 620–621.

MALEPEYRE (François).

Caoutchouc et gutta-percha.   (Manuel Roret.)   Paris, 1855. 12°.

MARCHAND (Louis).

Une mission forestière en Antriehe. Arbois, 1869. 8°.

MAUNY DE MORNAY (—.).

Livre du forestier. Guide complet de la culture, de l'exploitation des bois, *etc.* (Manuel Roret.) Paris, 1838. 18°. Ed. 2, 1842.

MORIÈRE (Jules).

Resumé des conférences agricoles sur la culture, le rouissage et le teillage du lin. Caen, 1864. 16°.

NAIRONE (Antonio Fausto).

Discorso della saluifera bevanda Cahve, ó vero Café . . . trasportato dalla Latina, alla lingua Italiana da Er. Frederic. Vegilin, *etc.* Roma, 1671. 12°.

NEUFVILLE (W. de).

Cinchona bark for the Pharmacopoeia. *Pharm. Journ.* III. xii. (1881) 369–370.

NEWTON (George William).

A treatise on the growth and future management of Timber Trees, *etc.* London, 1859. 8°.

PETERSEN (Franz).

De potv Coffi. Francofvrti, 1766. 4°.

Published under the initials F. P. only.

POULAIN (H.).

Production des coton dans nos colonies. Paris, 1863. 8°.

PUTON (Alfred).

L'aménagement des foréts. Traité pratique de la conduite des explorations des foréts, *etc.* Paris, 1867. 8°. Ed. 2. 1874. 18°.

QᴜᴇLUS (D.). *See* CHELUS (—. de).

RAMBALDI (Angelo).

Ambrosia arabica overo della salutare bevanda Café. Bologna, 1691. 12°.

REYBAUD (Louis).

Le coton, son régime, ses problèmes, son influence en Europe, *etc.* Paris, 1863. 8°.

REYNOSO (Alvaro).

Notas acercas del cultiva en camellones; agricultura de los indigenas de Cuba y Haiti. Paris, 1881. 8°.

Romano (G.).

Il frumento quale foraggio.　Milano, 1881.　8°.

Rousset (Antonin).

Recherches expérimentales sur les écorces à tan du chêne yeuse, relativément à la production et à l'aménagement des foréts de cette essence.　Paris, 1878.　4°.

Saint-Felix (A. J. M. de).

Instruction pratique sur la culture forestiére dans les terres fortes ou argileuses du midi.　Toulouse, 1841.　12°.

Schenk (Ernst).

Atlas der vorzueglichsten Handelspflanzen zu Schwarzkopf's Lehrbuch der Colonial- und Spezerei-Waarenkunde.　Mit erlaeuterndem Texte, *etc.*　Jena, [1853–54].　4°.

Schroff (Karl D.).

Lehrbuch der Pharmacognosie.　Wien, 1853.　8°.

Lehrbuch der Pharmacologie mit besonderer Beruecksichtung der oesterreichischen Pharmacopoe vom Jahre 1855.　Wien, 1856. 8°.

Das pharmacologische Institut der Wiener Universitaet, *etc.*　Wien, 1865.　8°.

Schwarzkopf (S. A.).

Lehrbuch der Droguenwaarenkunde.　Leipzig, 1855.　8°.

—— Lehrbuch der Colonial- und Spezerei-Waarenkunde.　Ed. 2. Jena, 1858.　8°.

Sebert (Hippolyte).

Notice sur les bois de Nouvelle Calédonie suivie de considèrations générales sur les propriétés mécaniques des bois, *etc.*　Paris, [1874].　8°.

Silva Coutinho (Joao Martino da).

Gommas, resinas, e gommas-resinas (in Relatio sorie a Exposiçao universal de 1867, redigido pelo secretairo da Commissão brazil-eira.　Tome ii. pp. 244, *etc.*)　Paris, 1868.　8°.

Souviron (A. R.).

De la culture du lin en Algèrie, de ses avantages et de l'utilité de son introduction dans l'assolement des terrains non arrosables. Paris, 1860.　8°.

Sylvestre Dufour (Philippe).

Traité nouveux et curieux du café, du thé, et du chocolate, *etc.* Ed. 2. Lyon, 1688.　12°.

SYLVESTRE DUFOUR (Philippe) *continued :*—

—— Traité . . . A quoy on a adjouté dans cette édition, la meilleure de toutes les methodes, qui manquoit à ce livre, pour composer l'excellent chocolate, par Mr. St. Disdier. Ed. 3. La Haye, 1693. 12°.

> This author's name has been previously given on page 55 as DUFOUR, on the authority of Quérard, La France litteraire, vol. ii. p. 642.

TERRACCIANO (Nicola).

La Peronospora viticola De Bary. Caserta, 1881. 8°.

THOMAS (Jean Basile).

Traité général de statistique, culture et exploitation des bois. Paris, 1840. 2 vols. 8°.

VALLIER (J.).

Petit manuel du planteur de coton, *etc.* Paris, 1862. 8°.

VATTEMARE (Hippolyte).

Le Fibrilia, substitut pratique et économique du coton; traité comprenant la description compléte du procédé de cotonisation du lin, du chanvre, du jute, de l'herbe de Chine et des autres fibres de même nature. Traduit de l'americain par H. V., *etc.* Paris, 1861. 8°.

VÉRET (Benjamin).

Le lin et sa culture. Paris, 1866. 8°.

VIDAL (—.).

Guide pratique à l'usage des gardes forestiers, traitant des arbres et arbustes forestiers, de l'enseignement des diverses espèces et de l'agriculture forestière, suivie d'un dictionnaire forestier raisonné, *etc.* Versailles, 1861. 8°.

———

Tobacco culture. Practical details, from the selection and pre-paration of the seed and the soil, to harvesting, curing and marketing the crop . . . Plain directions as given by fourteen experienced cultivators, residing in different parts of the United States, *etc.* New York, 1863. 8°.

VEGETABLE TECHNOLOGY.

PART II.

INDEX.

CORRIGENDA.

Pages 15 to 17, BENTLEY *should follow* BENOIT.
Page 18, *line* 13, *after* Ravenna., *insert* Ravenna,.
,,   19 ,,   4, *for* J. *read* R. J.
,,   20 ,,   16, *delete* C. M.
,,   60, *last line but one, insert* Carácas, 1874.   8°.
,,   75, *line* 22, *for* GROSS (G.) *read* GROSS (Heinrich).
,,   78 ,,   10 ,, Curcos *read* Curcas.
,,   78 ,,   12 ,, Mainhot *read* Manihot.
,,   119 ,,   36 ,, Krapf *read* Krapp.
,,   124 ,, 18 ,, qulques *read* quelques.
,,   143 ,,   3 ,, et sulla *read* e sulla.
,,   155 ,,   12 ,, SCAMALTZ *read* SCHMALTZ.
,,   158 ,,   14 ,, 18   *read* 1879.
,,   165 ,,   10 ,, Wilkomm *read* Willkomm.
,,   170, *lines* 5 *and* 10, *for* et *read* e.
,,   172, SCHINDLER *should precede* SCHLAGINTWEIT.
,,   174, *line* 17, *for* Standen *read* Stauden.
,,   175, SCHWEITZER *should precede* SCHWEIZER.
,,   183, *line* 37, *for* Auguste *read* Augustin.
,,   184 ,,   1, *after* (J. Léon) *insert* & Augustin DELONDRE.
,,   185 ,,   14, *for* 1865 *read* 1685.
,,   187 ,,   5 *from bottom, for* 1880, *read* 1878.
,,   189 ,,   3, *delete the period after* et.
,,   198 ,,   31, *for* Calédoine *read* Calédonie.
,,   208 ,,   21 ,, Ruecksight *read* Ruecksicht.
,,   208 ,,   36 ,, Lund- *read* Land-.
,,   209 ,, 21 ,, von wendeten *read* verwendeten.

STEPHEN AUSTIN AND SONS, PRINTERS, HERTFORD.